ALL THE UNIVERSE!

FASTER THAN LIGHT TACHYON QUARK STARSHIPS & PARTICLE ACCELERATORS WITH THE LHC AS A PROTOTYPE STARSHIP DRIVE

Stephen Blaha, Ph.D.

Pingree–Hill Publishing

ISBN: 978-0-9845530-1-3

Cover Credits
Cover by Stephen Blaha © 2011. Starry background used with the kind permission of NASA. Background LHC Photograph/Diagram by Christiane Lefèvre, © CERN 2008, used with the kind permission of CERN, Geneva, Switzerland.

May 11, 2011
rev. 00/00/01

To the Linear Hadron Collider (LHC) at
CERN:
Starship Test Prototype
Opening the Door to
Higher Energies
And the Stars

Some Other Books by Stephen Blaha

The Standard Model's Form Derived From Operator Logic, Superluminal Transformations And GL(16) (ISBN: 978-0-9845530-2-0, Pingree-Hill Publishing, Auburn, NH, 2010)

The Algebra of Thought & Reality: The Mathematical Basis for Plato's Theory of Ideas, and Reality Extended to Include A Priori Observers and Space-Time; Second Edition (ISBN: 9780981904931, Pingree-Hill Publishing, Auburn, NH, 2009)

The Metatheory of Physics Theories, and the Theory of Everything as a Quantum Computer Language (ISBN: 097469584X, Pingree-Hill Publishing, Auburn, NH, 2005)

The Origin of the Standard Model: The Genesis of Four Quark and Lepton Species, Parity Violation, the ElectroWeak Sector, Color SU(3), Three Visible Generations of Fermions, and One Generation of Dark Matter with Dark Energy (ISBN: 0974695882, Pingree-Hill Publishing, Auburn, NH, 2007)

Physics Beyond the Light Barrier: The Source of Parity Violation, Tachyons, and A Derivation of Standard Model Features (ISBN: 0974695874, Pingree-Hill Publishing, Auburn, NH, 2007)

Quantum Theory of the Third Kind: A New Type of Divergence-free Quantum Field Theory Supporting a Unified Standard Model of Elementary Particles and Quantum Gravity based on a New Method in the Calculus of Variations (ISBN: 0974695831, Pingree-Hill Publishing, Auburn, NH, 2005)

Quantum Big Bang Cosmology: Complex Space-time General Relativity, Quantum Coordinates™ Dodecahedral Universe, Inflation, and New Spin 0, ½, 1 & 2 Tachyons & Imagyons™ (ISBN: 0974695815, Pingree-Hill Publishing, Auburn, NH, 2004)

SuperCivilizations: Civilizations as Superorganisms (ISBN: 978-0-9819049-8-6, McMann-Fisher Publishing, Auburn, NH, 2010)

Available on bn.com, Amazon.com, and other web sites as well as at better bookstores (through Ingram Distributors).

Preface

The faster-than-light starship proposed in this book is contingent on quarks having complex-valued spatial momenta. This author has recently shown (Blaha (2010a)) that a theory built on this assumption yields the Standard Model of Elementary Particles exactly as it is current known experimentally – both the fermion and vector boson spectrum of particles, and SU(3) $\otimes$SU(2) $\otimes$U(1) – plus an additional U(1) symmetry and WIMPs. On this basis it appears that one has good reason to believe that quarks have complex 3-momenta.

This book is a design study for a superluminal starship, driven by thrust generated by tachyonic quark particle accelerators, capable of reaching the stars in this galaxy, and ultimately in other galaxies, in short periods of earth time ranging from months to a few years. It is a further development of the starship proposal described in the book *To Far Stars and Galaxies* by this author with a new, simpler starship engine design, and additional detail on starship requirements for suspended animation and nuclear engines as well as construction requirements.

It is the author's hope that an international consortium of space-faring nations can build a working prototype of this starship and then build the starship itself over, perhaps, a thirty-year period. Unlike other starship proposals, there does not appear to be any showstopper in the creation and use of this starship although there are major technical issues to be resolved in its construction.

At the time of this writing the United States appears headed towards a deficit of ten or eleven trillion dollars over the next ten years. It appears that a likely cost for starship R&D would be of the order of half a trillion dollars spread over perhaps thirty years – a small amount relative to the projected US deficit – with an enormous reward for success. NASA has recently proposed a 100 year starship program. We believe that the starship project can proceed far more rapidly by leveraging the investment in the Large Hadron Collider (LHC) at CERN by modifying it to be a starship engine prototype after it has performed its scheduled elementary particle experiments in the next 6 – 8 years. The enormous expenditure in constructing the 27 km LHC tunnel and other supporting accelerator tunnels can be recovered by the conversion of the LHC to a starship prototype with superluminal drive, and testing starship engine concepts and performance. We suggest a reasonable approach to beginning this effort.

The starship venture is of great potential significance to the future of Mankind. The cost is relatively small; the benefits are potentially enormous.

CONTENTS

FIGURES

1. Superluminal Starship Program

In 2008 and 2009 this author wrote two books proposing the research and design of a starship capable of journeys to other stars and galaxies in short periods of time up to a couple of earth months for stars and up to a couple of years to other galaxies. The travel times envisioned would be comparable to travel times around the world in sailing ships in the 17th century. For example a starship traveling at 5,000c (five thousand times the speed of light) could reach the center of the Milky Way (a distance of 30,000 light years) in six earth years. Reaching the nearest stars (about 4 light years) would take about 1/3 of a day!

In 2009 in the second book[1] *To Far Stars and Galaxies* a series of steps was described for the starship design and implementation. Since the writing of this book a number of very significant developments have occurred that have significantly furthered the prospects of this project:

1.1 Quark Drive Progress:

This author has made a detailed study of the form of the Standard Model of Elementary Particles and shown its form can derived from space-time considerations.[2] An important aspect of the derivation is that d quarks – a type of quark in protons and neutrons – are tachyons although their binding within protons and neutrons masks (hides) their tachyonic nature. Tachyons are particles that travel faster than the speed of light. If the d quarks can be freed from protons and neutrons, and accelerated they can provide a tachyonic thrust to power starships faster than the speed of light. The success in obtaining the Standard Model's form including the $SU(2) \otimes U(1)$ ElectroWeak symmetry and the Strong Interactions $SU(3)$ symmetry from the superluminal (faster than light) Lorentz transformation sector strongly supports the concept of tachyonic (superluminal) d quarks. These quarks can be produced in Hadron Colliders such as the LHC currently running at CERN in Switzerland.

---

[1] Chapter 11.
[2] Blaha (2010a).

Much of the remainder of this book is devoted to a preliminary design study of d quark production and acceleration to produce superluminal thrust. It is followed by a proposal to modify the LHC to be a starship engine prototype after its elementary particle physics research phase is completed in 6 – 8 years. A successful prototype, after extensive testing, would be followed by the construction of a starship in orbit around the earth.

1.2 Planet Search Phase

A plethora of planets, including many earth size planets circling nearby stars, have been found in initial efforts of NASA's Kepler Space Telescope: 1,235 planets including 54 earth-size planets in the habitable zone of their stars. (as of February 2, 2011)

These planets (and more that are likely to be found) offer the possibility of earth-like environments suitable for human colonization – more so than the other barren, inhospitable planets and moons of our solar system. Thus many possible destinations for a starship are now very likely to exist.

If high-speed starships are developed then the earth-like planets of other stars may be better choices for colonization than the planets of our solar system.

1.3 Quark-Gluon Plasmas/Perfect Fluids

The RHIC, which has successfully created interaction regions containing quark-gluon plasmas, was among the first of a series of increasingly powerful, circular accelerators that produce quark-gluon plasmas and perfect fluids.

The CERN LHC has started producing quark-gluon perfect fluid fireballs[3] in order to study their characteristics. Later in this book we will consider superluminal quark perfect fluids that can be focussed by laser or particle beams to produce superluminal thrust for starships.

---

[3] These plasmas and perfect fluids are produced by the collision of heavy ions of lead and gold at high energy

1.4 Design and Development of Plasma/Perfect Fluid Thrusters

The successful development of stable, high energy, high quark density quark-gluon thrust is the key to quark drive. Starship thrust must be complex in nature to achieve the high superluminal speeds.

We discuss the rings and accelerator configuration to produce superluminal thrust in the succeeding chapters.

1.5 Nuclear or Fusion Engine Phase

In addition our Blaha (2009b) proposal required nuclear or fusion engines for local movement within a star system. Since Blaha (2009b) was written, Russia has announced a major program to develop nuclear rocket propulsion. Presumably Russian nuclear engines, or American equivalents,[4] will be available for "short distance" starship propulsion within star systems.

1.6 Long Term Suspended Animation

As described in Blaha (2009b) biomedical research must develop a means of long term suspended animation with a minimal initial goal of 10 years and an optimal goal of 1,000 years. Starship occupants would be put in suspended animation during trips to counteract the effects of time dilation.[5]

Occupants would sleep for many years during the trip so that upon their return to earth their biological clocks would be roughly in sync with the elapsed earth time.

At the moment studies of hibernation and suspended animation are in an early stage. However, for trips to nearby stars suspended animation would not be required.

1.7 Starship Prototype and Test Phase

After the R & D phase a "small" prototype for test purposes should be constructed and tested on earth (at the Large Hadron Collider

---

[4] Preliminary American designs exist from NASA work in this area in the 1950's and 1960's.

[5] In starships traveling much faster than the speed of light time will pass much more quickly than on earth. Thus, for example, a trip to Proxima Centauri (about four light years away) at 5,000c would take about 1/3 of a day in earth time but about $5000\times(1/3) = 4$ years of time in the starship. This hitherto unremarked phenomena of superluminal time dilation is discussed in Blaha (2010a) and his earlier books.

(LHC) at CERN?) and in space. A study of test results and detailed design variations will undoubtedly lead to significant improvements both in performance and safety. An example of performance improvements that can be obtained through careful analysis of prototypes is computer chip fabrication in the 1980s. United States chip manufacturers were producing computer chips of which only 20% were free of defects. The result was a relatively high manufacturing cost of US made chips. Japanese manufacturers in the early 1980s were producing chips of which 80% were free of defects. As a result they could easily undersell American manufacturers. How did Japanese manufacturers learn to produce high quality chips? They carefully observed each stage of the chip manufacturing process and paid particular attention to the feedback/suggestions of workers on the manufacturing lines on improving chip manufacturing.

This same approach of using an interactive feedback approach in building the starship prototype should lead to a starship with superior performance characteristics. We propose the use of a modified LHC to create the first starship engine (at relatively low cost) in a subsequent chapter.

1.8 A Working Starship Ready to Explore

After developing and testing a successful prototype, then a full-scale starship should be built in space in orbit around the earth. And then it's off to the stars!

1.9 Concluding Comments

Blaha (2009b), and this book, outline an ambitious program to conquer the solar system and the universe that ultimately would include to a massive effort to move people from earth to other earth-like planets similar to the massive transfer of people from Europe to the Americas in past centuries. The only serious long-term option for Mankind is to make major space efforts in the Solar System, then the galaxy, and then the universe so the universe will *not* be "wasted" space.

The initial price is high – perhaps half a trillion dollars – but doable. The price for not acting, and confining mankind to an overcrowded, and obviously declining planet, is much higher.

2. Tachyons and Superluminal Transformations

2.1 Tachyon Starship Propulsion

In this book we will consider the use of d-quark tachyons as starship propellant. This new approach appears to be feasible using extensions of current technology and does not require massive resources beyond those currently available unlike other starship proposals. In the following sections we will describe some important features of superluminal Lorentz transformations – the key to tachyonic motion and ultimately to tachyon starship drive. In subsequent chapters we will describe a starship drive based on tachyon thrust. Tachyon acceleration to create superluminal thrust is very different from subluminal particle acceleration as done today in various accelerators around the world.

Tachyons,[6] particles traveling at a speed faster than light, were first proposed[7] in the 1960's. The concept of a superluminal particle was based on the possibility that the speed of light is not an absolute limit but a type of "horizon" beyond which sub-light particles could not go in our normal, almost flat space-time. The concept of superluminal particles was recognized as not creating paradoxes but early work on the quantum theory of tachyons was not able to establish a successful free quantum field theory of tachyons. Blaha (2007a) developed the first successful quantum theory of free tachyons.

Einstein said that he was led to the Special Theory of Relativity by considering an observer traveling at the speed of light and viewing an electromagnetic wave traveling at the same speed. The electromagnetic wave would appear to be static – not having the oscillatory nature of a wave – contrary to the theory of electromagnetism. He concluded that no

---

[6] Also called superluminal particles.
[7] S. Tanaka, Prog. Theoret. Phys. (Kyoto) **24**, 171 (1960); O. M. P. Bilaniuk, V. K. Deshpanda and E. C. G.; Sudarshan, Am J. Phys. **30**, 718 (1962); G. Feinberg, Phys. Rev. **159**, 1089 (1967); M. E. Arons & E. C. G. Sudarshan, Phys. Rev. **173**, 1622 (1968). Feinberg coined the term tachyon in his 1967 paper.

observer (or particle that might "hold" an observer) could travel at the speed of light and thus all particles must travel at a speed below the speed of light. In fact, he actually proved that particles with mass cannot travel at exactly the speed of light.

He did not consider the possibility that particles might exist that travel at a speed greater than the speed of light. Blaha (2007a) considered exactly that case and found that particles traveling faster than the speed of light would "see" electromagnetic waves traveling at precisely the speed of light and oscillate according to the laws of electromagnetism.[8] Thus Einstein's thought experiment *only* ruled out the possibility of massive particles traveling at exactly the speed of light.

We will therefore assume tachyons can exist in nature but remain to be found experimentally. In previous books we showed that 1) particles entering a black hole acquire a superluminal speed and become tachyons, and 2) that one can derive the Standard Model of Elementary Particles if one assumes neutrinos and down-type quarks such as the d quark are tachyons.[9] Since neutrinos have an extremely small mass it is not yet possible to verify that they are tachyons. Since quarks are confined within protons, neutrons, and so on, the nature of down-type quarks is also not known with certainty. However the difficulties of interpretation of the spin characteristics of deep inelastic electron-nucleon scattering suggest quarks are not normal spin ½ particles. These discrepancies may be explainable if down-type quarks are tachyons. Tachyons have different spin characteristics than normal spin ½ particles.

Given the possibility of tachyons in nature, and remembering the history of black holes which were predicted many years ago and only found recently, it is reasonable to expect that tachyons will be found – perhaps in studies of the quark-gluon plasmas that are now being produced at new accelerators in Au-Au and other heavy ion scattering. Since particles inside black holes are tachyons, the question devolves to finding tachyons outside of black holes.

---

[8] Blaha (2007a) p. 12 and earlier work.
[9] Blaha (2006), (2007a), (2007b), (2008), (2010a).

2.2 Superluminal Lorentz Transformations – The Extension of Special Relativity to Include Faster-than-Light Particles - Tachyons

This section can be skipped by the non-technical reader. The Lorentz group, the essence of Special Relativity, relates the coordinates of an event in two coordinate systems that differ by a relative velocity whose magnitude is less than the speed of light. We imagine an observer on one "particle" (the "lab" observer) observing at event at time t at the position (x, y, z). Another observer on another "particle" traveling at a speed v in the x direction relative to the first observer (as depicted in Fig. 2.1) observes the same event at time t' at the position (x', y', z'). The relation between the coordinates of the two observers is given by the Lorentz transformation

$$t' = \gamma(t - \beta x/c) \qquad (2.1)$$
$$x' = \gamma(x - \beta ct)$$
$$y' = y$$
$$z' = z$$

or, in matrix form,

$$X' = \Lambda_L(\omega, \mathbf{u} = (1,0,0))X$$

where $\Lambda_L(\omega, \mathbf{u} = (1,0,0))$ is a matrix representation of the transformation in the x direction symbolized by $\mathbf{u} = (1,0,0)$, where $\beta = v/c$, c is the speed of light, and where $\gamma = (1 - \beta^2)^{-\frac{1}{2}}$.

If β is less than one (sublight speed) then the coordinates of an event are related by eq. 2.1, and specify the time and location of an event from the viewpoint of an observer at rest in each coordinate system. If β is greater than one (superluminal speed) then the coordinates of an event are still related by eq. 2.1 but now t' and x' are imaginary numbers.

How can we physically understand this state of affairs? Well, if we consider the realities of the observer in the "primed" coordinate system it is clear he/she will measure the x' distance with a ruler that measures real numbers, and he/she will measure time t' with a clock that measures real numbers. So the imaginary values of x' and t' in eq. 2.1 only appear in the relation between the coordinate systems. If we denote the actual values measured by the primed coordinate system observer as t_r'' and x_r'' then eq. 2.1 for a superluminal relative speed v becomes

$$t_r'' = i\gamma(t - \beta x/c) \qquad (2.2)$$
$$x_r'' = i\gamma(x - \beta ct)$$
$$y' = y$$
$$z' = z$$

where $\beta > 1$.

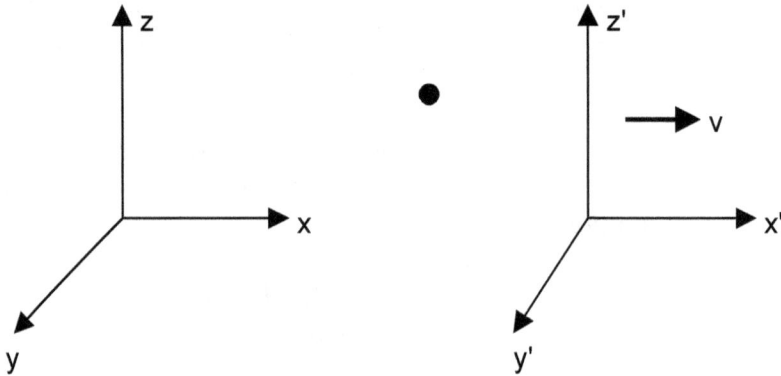

Figure 2.1. Two coordinate systems having a relative speed v in the x direction. The black circle represents an event. The unprimed coordinate system is the "lab" system. The "primed" coordinate system is the system of an observer on a particle moving with speed v along the parallel x and x' axes.

The 4-vector inner product is invariant for $\beta > 1$ for t' and x' as given by eq. 2.1:

$$c^2t^2 - x^2 - y^2 - z^2 = c^2t'^2 - x'^2 - y'^2 - z'^2 \qquad (2.3)$$

just as it is for $\beta < 1$. (Note eq. 2.3 for double primed coordinates is

$$c^2t^2 - x^2 - y^2 - z^2 = -c^2t''^2 + x''^2 - y'^2 - z'^2 \qquad (2.3a)$$

The physical interpretation of complex coordinates generated by a superluminal transformation is straightforward. Since a superluminal transformation is a linear transformation a straight line whose points are specified by real numbers will map, in general, into a straight line whose points are specified by complex numbers. Thus each point on the lab straight line (a_x, a_y, a_z) is mapped by a superluminal transformation to a three dimensional point whose three coordinate values are complex numbers $(a_{xr}' + ia_{xi}', a_{yr}' + ia_{yi}', a_{zr}' + ia_{zi}')$. An observer in the primed system will, of course, measure real numbers with a ruler between points. These real number coordinate values can be determined by introducing a new transformation that maps the complex numbers generated by superluminal transformations into the real numbers that the primed system observer would measure.

This transformation is of great significance because it leads to a hitherto unstated $SU(2) \otimes U(1) \oplus U(1)$ symmetry. The $SU(2) \otimes U(1)$ part of this new symmetry can be identified as the source of the $SU(2) \otimes U(1)$ symmetry of the ElectroWeak sector of The Standard Model. (See Blaha (2010a).) The additional $U(1)$ symmetry is the source of WIMPs (Weakly Interacting Massive Particles).

We introduce this new transformation by reconsidering the previous simple example (Fig. 2.1) wherein one coordinate system is traveling at a speed v in the x direction with respect to the "laboratory" system. The coordinates in the two reference frames are related by eq. 2.1. We now define a transformation that maps the real coordinates of the unprimed reference frame to real coordinates in the primed reference frame.

$$\Pi_L(\mathbf{u}) = \begin{bmatrix} -i & 0 & 0 & 0 \\ 0 & -i & 0 & 0 \\ 0 & 0 & 1 & 0 \\ 0 & 0 & 0 & 1 \end{bmatrix} \tag{2.4}$$

where $\mathbf{u}$ is the unit vector corresponding to the direction of $\mathbf{v}$ (the positive x direction in this example). Using $\Pi_L(\mathbf{u})$ in eq. 2.1 we obtain an overall transformation from real coordinates to real coordinates:

$$X'' = \Pi_L(\mathbf{u})X' = \Pi_L(\mathbf{u})\Lambda_L(\omega, \mathbf{u} = (1,0,0))X$$

or

$$\begin{aligned} t'' &= \gamma_s(t - \beta x) \\ x'' &= \gamma_s(x - \beta t) \\ y'' &= y \\ z'' &= z \end{aligned} \tag{2.5}$$

where $\gamma_s = i\gamma$.

An observer in the primed reference frame would consider his/her time to be real when measured on a clock, and his/her distances along the x-axis to be real when measured with a ruler. Thus eq. 2.5 makes good sense physically because in any reference frame observers measure real distances and real times. For this reason we will call transformations of the type of eq. 2.5 – from real coordinates to real coordinates – *physical* superluminal transformations.

This simple example generalizes to arbitrary relative velocities **v**. First we note that the Lorentz transformation for a velocity **v** that is a rotation of the velocity in the x-direction (**v** = |**v**|R**u** where R is the relevant rotation matrix) has the form

$$\Lambda_L(\omega, \mathbf{v}) = \mathcal{R}(\mathbf{v}/v, \mathbf{u})\Lambda_L(\omega, \mathbf{u} = (1,0,0))\mathcal{R}^{-1}(\mathbf{v}/v, \mathbf{u}) \qquad (2.6)$$

where $\mathcal{R}(\mathbf{v}/v, \mathbf{u})$ is a rotation from the velocity direction **u** to direction **v**/v.

The original transformation (eq. 2.5) can be written as

$$\Pi_L(\mathbf{u})\Lambda_L(\omega, \mathbf{u} = (1,0,0)) = \Pi_L(\mathbf{u})\mathcal{R}^{-1}(\mathbf{v}/v, \mathbf{u})\Lambda_L(\omega, \mathbf{v})\mathcal{R}(\mathbf{v}/v, \mathbf{u}) \qquad (2.7)$$

Consequently the combined transformation for velocity **v** is

$$\begin{aligned}
\mathcal{R}(\mathbf{v}/v, \mathbf{u})\Pi_L(\mathbf{u})&\Lambda_L(\omega, \mathbf{u} = (1,0,0))\mathcal{R}^{-1}(\mathbf{v}/v, \mathbf{u}) \\
&= \mathcal{R}(\mathbf{v}/v, \mathbf{u})\Pi_L(\mathbf{u})\mathcal{R}^{-1}(\mathbf{v}/v, \mathbf{u})\Lambda_L(\omega, \mathbf{v}) \\
&= \Pi_L(\mathbf{v}/v)\Lambda_L(\omega, \mathbf{v}) \qquad (2.8)
\end{aligned}$$

Thus for a Lorentz transformation $\Lambda_L(\omega, \mathbf{v})$ for velocity **v** we see that we can define a subsidiary transformation $\Pi_L(\mathbf{v}/v)$ of the form

$$\Pi_L(\mathbf{v}/v) = \mathcal{R}(\mathbf{v}/v, \mathbf{u})\Pi_L(\mathbf{u})\mathcal{R}^{-1}(\mathbf{v}/v, \mathbf{u}) \qquad (2.9)$$

The general form of $\mathcal{R}(\mathbf{v}/v, \mathbf{u})$, is

$$\mathcal{R}(\mathbf{v}/v, \mathbf{u}) = \begin{bmatrix} 1 & 0 & 0 & 0 \\ 0 & & & \\ 0 & & \mathcal{R}_3(\mathbf{v}/v, \mathbf{u}) & \\ 0 & & & \end{bmatrix} \qquad (2.10)$$

where $\mathcal{R}_3(\mathbf{v}/v, \mathbf{u})$ is a 3×3 rotation matrix that can be expressed in terms of the generators of the 3-dimensional rotation group as

$$\mathcal{R}_3(\mathbf{v}/v, \mathbf{u}) = \exp(i\boldsymbol{\theta}\cdot\mathbf{J}) \qquad (2.11)$$

The rotation angles $\boldsymbol{\theta}$ are real numbers since we are rotating the real vector $\mathbf{u}$ to the real number $\mathbf{v}/v$. Given the form of eq. 2.11 then we see that the form of $\Pi_L(\mathbf{v}//v)$ is

$$\Pi_L(\mathbf{v}/v) = \begin{bmatrix} -i & 0 & 0 & 0 \\ 0 & & & \\ 0 & & \mathcal{R}_3(\mathbf{v}/v, \mathbf{u})\Pi_{L3}(\mathbf{u})\mathcal{R}_3^{-1}(\mathbf{v}/v, \mathbf{u}) \\ 0 & & & \end{bmatrix} \qquad (2.12)$$

where

$$\Pi_{L3}(\mathbf{u}) = \begin{bmatrix} -i & 0 & 0 \\ 0 & 1 & 0 \\ 0 & 0 & 1 \end{bmatrix} \qquad (2.13)$$

If we consider the case of an infinitesimal rotation $\boldsymbol{\theta}$ to first order in $\boldsymbol{\theta}$

$$\mathcal{R}_3(\mathbf{v}/v, \mathbf{u}) \simeq I + i\boldsymbol{\theta}\cdot\mathbf{J} \qquad (2.14)$$

then

$$\begin{aligned} \Pi_{L3}(\mathbf{v}/v) &= \mathcal{R}_3(\mathbf{v}/v, \mathbf{u})\Pi_{L3}(\mathbf{u})\mathcal{R}_3^{-1}(\mathbf{v}/v, \mathbf{u}) \\ &\simeq \Pi_{L3}(\mathbf{u}) + i\boldsymbol{\theta}\cdot\mathbf{J}\Pi_{L3}(\mathbf{u}) - i\Pi_{L3}(\mathbf{u})\boldsymbol{\theta}\cdot\mathbf{J} \\ &\simeq \Pi_{L3}(\mathbf{u})[I + i\Pi_{L3}^{-1}(\mathbf{u})[\boldsymbol{\theta}\cdot\mathbf{J}, \Pi_{L3}(\mathbf{u})] \end{aligned} \qquad (2.15)$$

where $\Pi_{L3}^{-1}(\mathbf{u})$ is the inverse of $\Pi_{L3}(\mathbf{u})$ and [...] represents the commutator. Thus for arbitrary rotations eq. 2.15 implies

$$\begin{aligned} \Pi_{L3}(\mathbf{v}/v) &= \mathcal{R}_3(\mathbf{v}/v, \mathbf{u})\Pi_{L3}(\mathbf{u})\mathcal{R}_3^{-1}(\mathbf{v}/v, \mathbf{u}) \\ &= \Pi_{L3}(\mathbf{u})\exp\{i\Pi_{L3}^{-1}(\mathbf{u})[\boldsymbol{\theta}\cdot\mathbf{J}, \Pi_{L3}(\mathbf{u})]\} \end{aligned} \qquad (2.16)$$

We can find the general form of $\Pi_{L3}(\mathbf{v}/v)$ by considering the case of eq. 2.4 in more detail. The exponentiated matrix expression in 2.16 can written

$$\Pi_{L3}^{-1}(\mathbf{u})[\boldsymbol{\theta}\cdot\mathbf{J}, \Pi_{L3}(\mathbf{u})] = \Pi_{L3}^{-1}(\mathbf{u})\boldsymbol{\theta}\cdot\mathbf{J}\Pi_{L3}(\mathbf{u}) - \boldsymbol{\theta}\cdot\mathbf{J}$$

$$= \boldsymbol{\theta} \cdot \mathbf{Q} \qquad (2.17)$$

where

$$\mathbf{Q} = \Pi_{L3}^{-1}(\mathbf{u})\mathbf{J}\Pi_{L3}(\mathbf{u}) - \mathbf{J} = \mathbf{Q}' - \mathbf{J} \qquad (2.18)$$

The matrices Q_i can be evaluated using eq. 2.13 and the matrix representations for the rotation generators J_i: which are equivalent to the SU(2) generators T_i:

$$J_1 = \begin{bmatrix} 0 & 0 & 0 \\ 0 & 0 & -i \\ 0 & i & 0 \end{bmatrix} = T_1 \qquad (2.19)$$

$$J_2 = \begin{bmatrix} 0 & 0 & i \\ 0 & 0 & 0 \\ -i & 0 & 0 \end{bmatrix} = T_2 \qquad (2.20)$$

$$J_3 = \begin{bmatrix} 0 & -i & 0 \\ i & 0 & 0 \\ 0 & 0 & 0 \end{bmatrix} = T_3 \qquad (2.21)$$

The rotation generators satisfy the commutation relations

$$[J_i, J_j] = i\epsilon_{ijk}J_k \qquad (2.22)$$

as do the SU(2) generators:

$$[T_i, T_j] = i\epsilon_{ijk}T_k \qquad (2.23)$$

We can calculate Q' from eqs. 2.13 and 2.16 – 2.20 and obtain

$$Q'_1 = \begin{bmatrix} 0 & 0 & 0 \\ 0 & 0 & -i \\ 0 & i & 0 \end{bmatrix} \qquad (2.24)$$

$$Q'_2 = \begin{bmatrix} 0 & 0 & -1 \\ 0 & 0 & 0 \\ -1 & 0 & 0 \end{bmatrix} \qquad (2.25)$$

$$Q'_3 = \begin{bmatrix} 0 & 1 & 0 \\ 1 & 0 & 0 \\ 0 & 0 & 0 \end{bmatrix} \qquad (2.26)$$

We note that each Q'_i is hermitean and the Q'_i satisfy the commutation relations:

$$[Q'_i, Q'_j] = i\epsilon_{ijk}Q'_k \qquad (2.27)$$

Consequently the set of Q'_i are also equivalent to SU(2) group generators. As a result the exponential factor in eq. 2.16:

$$\Pi_{L3}(\mathbf{v}/v) = \Pi_{L3}(\mathbf{u})\exp\{i\boldsymbol{\theta}\cdot(\mathbf{Q'} - \mathbf{J})\} \qquad (2.28)$$

is equivalent to a combination of SU(2) rotations not only in this case but in general for superluminal transformations. The factor $\Pi_{L3}(\mathbf{u})$ is not an SU(2) matrix since its determinant is not +1 but

$$\Pi'_{L3}(\mathbf{u}) = -i\Pi_{L3}(\mathbf{u}) \qquad (2.29)$$

is an SU(2) matrix since

$$\Pi'_{L3}{}^{-1}(\mathbf{u}) = \Pi'_{L3}{}^{\dagger}(\mathbf{u}) \qquad (2.30)$$
$$\det \Pi'_{L3}(\mathbf{u}) = 1 \qquad (2.31)$$

Thus

$$\Pi'_{L3}(\mathbf{v}/v) = \Pi'_{L3}(\mathbf{u})\exp\{i\boldsymbol{\theta}\cdot(\mathbf{Q'} - \mathbf{J})\} \qquad (2.32)$$

is an SU(2) rotation.

Thus the general form of superluminal transformation from a real set of coordinates to a real set of coordinates is[10]

$$\Pi_L(\mathbf{v}/v)\Lambda_L(\omega, \mathbf{v}) \qquad (2.33)$$

where

---

[10] The choice of the unit vector $\mathbf{u}$ and the angle vector $\boldsymbol{\theta}$ must be such that applying eq. 2.34 to a real set of coordinates yields a real set of coordinates.

$$\Pi_L(\mathbf{v}/v) = \begin{bmatrix} -i & 0 & 0 & 0 \\ 0 & & & \\ 0 & & \Pi_{L3}(\mathbf{u})\exp\{i\,\boldsymbol{\theta}\cdot(\mathbf{Q}' - \mathbf{J})\} & \\ 0 & & & \end{bmatrix} \qquad (2.34)$$

by eqs. 2.8, 2.12 and 2.29 – 2.32. The Lorentz condition for real to real transformations generalizes to

$$\Lambda(\mathbf{v})^T\Pi_L(\mathbf{v}/v)^\dagger G\Pi_L(\mathbf{v}/v)\Lambda(\mathbf{v}) = G \qquad (2.35)$$

Since superluminal transformations $\Lambda_L(\omega,\ \mathbf{v})$ transform real coordinates to complex coordinates in general, we can generalize the form of a superluminal transformation to

$$e^{i\varphi}\Pi_L(\mathbf{v}'/v')\Lambda_L(\omega,\ \mathbf{v}) \qquad (2.36)$$

where φ is a constant phase and $\mathbf{v}'$ is an arbitrary velocity. This generalization will satisfy the generalized Lorentz condition

$$\Lambda(\mathbf{v})^T\Pi_L(\mathbf{v}'/v')^\dagger e^{-i\varphi}G\ e^{i\varphi}\Pi_L(\mathbf{v}'/v')\Lambda(\mathbf{v}) = G \qquad (2.37)$$

but the transformation will, in general, yield a complex set of coordinates when applied to a set of real coordinates.

These considerations imply:

1. Any observer in a coordinate system will treat a complex 4-dimensional coordinate system as if it were a real 4-dimensional coordinate system with complex-valued straight lines along each dimension (assuming rectangular coordinates).

2. The transformation $e^{i\varphi}\Pi'_{L3}(\mathbf{v}/v)$ is a SU(2)⊗U(1) transformation that takes complex 3-dimensional spatial coordinates to complex 3-dimensional spatial coordinates. In particular straight lines map to straight lines.

3. Physical observations in the observer's coordinate system are invariant under SU(2)⊗U(1) rotations of the spatial coordinates and the multiplication of the time component by an arbitrary phase.

4. The matrix

$$\Pi'_L(\mathbf{v}/v, \chi, \varphi) = \begin{bmatrix} e^{i\chi} & 0 & 0 & 0 \\ 0 & & & \\ 0 & e^{i\varphi}\Pi'_{L3}(\mathbf{u})\exp\{i\,\boldsymbol{\theta}{\cdot}(\mathbf{Q'} - \mathbf{J})\} & & \\ 0 & & & \end{bmatrix} \qquad (2.38)$$

(where χ and φ are real numbers and **u** is a unit vector along any convenient coordinate axis) is an SU(2)⊗U(1)⊕U(1) transformation that transforms complex 4-dimensional coordinates to complex 4-dimensional coordinates. Note, $\Pi_L(\mathbf{v}/v) = \Pi'_L(\mathbf{v}/v, 3\pi/2, \pi/2)$ is a special case of $\Pi'_L(\mathbf{v}/v, \chi, \varphi)$. Due to the manifest form of eq. 2.38 we see

$$\Pi'_L{}^\mu{}_\alpha * \Pi'_L{}^\mu{}_\beta = [\Pi'_L{}^\dagger\Pi'_L]_{\alpha\beta} = I_{\alpha\beta} \qquad (2.39)$$

(with an implied sum over μ) or, in matrix form,

$$\Pi'_L{}^\dagger \Pi'_L = I \qquad (2.40)$$

and also[11]

$$\Pi'_L{}^\dagger G\Pi'_L = G \qquad (2.41)$$

5. Complex coordinate values of the type generated by superluminal transformations are transformable to real coordinates by a transformation of the form of eq. 2.38 – an SU(2)⊗U(1)⊕U(1) transformation. The complex coordinates are thus physically equivalent to corresponding real coordinate values in the sense that an observer in that frame would automatically use the real coordinates so obtained since rulers and clocks always measure real spatial coordinates and times.

---

[11] Eq. 2.35 is close to the defining condition for a Lorentz group element but the presence of complex cojugation rather than a transpose means Π'_L is outside the real and complex Lorentz groups.

6. The complex coordinates of any point obtained through a superluminal transformation can be transformed to a real set of coordinates by a transformation of the form of eq. 2.38.

SU(2)⊗U(1)⊕U(1) invariance can be shown to lead to SU(2)⊗U(1)⊕U(1) ElectroWeak sector of the Standard Model with WIMPs which can be restricted to an SU(2)⊗U(1)). See Blaha (2010a).

7. The form of eq. 2.37 implies that the space-time group of the Standard Model is

$$SU(2)\otimes U(1)\oplus U(1)\otimes L_c \qquad (2.37a)$$

where L_c is the complex Lorentz group. This group is a subgroup of the complex group GL(4). Space-time, in general has 4 complex dimensions. These dimensions can be treated as real dimensions because of the equivalence of complex coordinates to real coordinates under SU(2)⊗U(1)⊕U(1) transformations.

2.3 Time and Space Contraction and Dilation

This section can be "skim" read by the non-technical reader. In ordinary Lorentz transformations a moving ruler will appear to be shorter in the direction of its motion when measured in another reference frame. This phenomenon is called *Lorentz contraction.* In ordinary Lorentz transformations time intervals will appear to be longer when measured in another reference frame. This phenomenon is called *time dilation.*

In superluminal transformations contraction and dilation are more complicated as we will see in this section.

2.3.1 Superluminal Length Dilation/Contraction

In the case of a superluminal transformation we find *superluminal length contraction or dilation* can occur depending on the relative velocity. Consider the case of the transformation of eq. 2.1 above, which relates the primed reference frame traveling at speed v in the positive x direction to the unprimed reference frame. A ruler perpendicular to the x-axis will have the same length in both reference frames if its endpoints are simultaneously measured – perhaps by photographing it. The y and z equations in eqs. 2.1 specify this fact.

If the ruler is at rest in the primed reference frame and parallel to the x' axis, then a simultaneous measurement of its endpoints at the same time t_0 by an observer in the unprimed reference frame (perhaps by photographing it) will reveal either *length contraction and dilation* depending on the value of β. If the length is $L' = x'_2 - x'_1$ in the primed frame and $L = x_2 - x_1$ in the unprimed frame, then the equations:

$$x'_1 = \gamma_s(x_1 - \beta ct_0) \qquad (2.42)$$
$$x'_2 = \gamma_s(x_2 - \beta ct_0) \qquad (2.43)$$

where $\gamma_s = i\gamma$ imply

$$L' = \gamma_s L = (\beta^2 - 1)^{-\frac{1}{2}} L \qquad (2.44)$$

Thus we have three cases:

Case 1: $\beta \in <1, \sqrt{2}>$: $\qquad$ L < L' $\qquad$ Contraction $\qquad$ (2.45)

Case 2: $\beta = \sqrt{2}$: $\qquad$ L = L' $\qquad$ Equality $\qquad$ (2.46)

Case 3: $\beta \in <\sqrt{2}, \infty>$: $\qquad$ L > L' $\qquad$ Dilation $\qquad$ (2.47)

Thus β = v/c = √2 marks the point of change from the Lorentz contraction of lengths to dilation of lengths. This feature of superluminal motion, first noted in Blaha (2007a), has no counterpart in sublight motion.

The effect of the change at β = √2 on a starship is startling. Imagine the primed coordinate system is that of the starship and the unprimed coordinate system is that of the earth. (The motion of the earth is small and can be neglected relative to the high speed of the starship.) If the starship velocity moving on a straight line away from the earth is such that β is between 1 and √2 then a yardstick on the starship will appear to be shorter in the earth coordinate system – Lorentz contraction. If the starship velocity is such that β is greater than √2 then a yardstick on the starship will appear to be longer in the earth coordinate system – dilation. More interestingly, if the starship travels 1 light year in its coordinate system it will actually have traveled more than 1 light year in the earth's coordinate system – possibly *much* more than 1 light year in the earth's coordinate system – if the starship's speed has a β that is much greater than √2.

2.3.2 Superluminal Time Contraction/Dilation

In the case of a superluminal transformation *superluminal time contraction* is a possibility.[12] Consider again the case of the transformation of eq. 2.1 relating the primed reference frame traveling at speed v in the positive x direction to the unprimed reference frame. Consider the time interval between two events occurring at the same point x'_0 in the primed reference frame. From the viewpoint of an observer in the unprimed frame the events take place at different points x_1 and x_2. If the time interval is $T' = t'_2 - t'_1$ in the primed frame and $T = t_2 - t_1$ in the unprimed frame, then the inverse transformation to eq. 2.1 gives:

$$t_1 = \gamma_s(t'_1 + \beta x'_0/c) \qquad (2.48)$$
$$t_2 = \gamma_s(t'_2 + \beta x'_0/c) \qquad (2.49)$$

and implies

$$T = \gamma_s T' = (\beta^2 - 1)^{-\frac{1}{2}} T' \qquad (2.50)$$

Again there are three cases:

Case 1: $\beta \in <1, \sqrt{2}>$: $\qquad T > T' \qquad$ Dilation $\qquad (2.51)$

Case 2: $\beta = \sqrt{2}$: $\qquad T = T' \qquad$ Equality $\qquad (2.52)$

Case 3: $\beta \in <\sqrt{2}, \infty>$: $\qquad T < T' \qquad$ Contraction $\qquad (2.53)$

The time interval in the unprimed (earth) frame can be less than, equal to, or greater than the time interval in the primed frame when the events take place at the same spatial point.

Thus superluminal transformations are more complex than sublight Lorentz transformations (which only have time dilation) with respect to time dilation and contraction.

The effect of the time change at $\beta = \sqrt{2}$ on a starship is also startling. Again imagine the primed coordinate system is that of the starship and the unprimed coordinate system is that of the earth. If the starship velocity moving on a straight line away from the earth is such that β is between 1 and $\sqrt{2}$ then a time interval on the starship will appear

---

[12] Blaha (2007a).

to be longer in the earth coordinate system – time dilation. Or, stating it otherwise, time intervals on the starship will appear to be shorter than they appear on earth. A person on a starship would thus age more slowly from the point of view of a person on earth.

But if the starship velocity is such that β is greater than $\sqrt{2}$ then a time interval on the starship will appear to be shorter in the earth coordinate system – time contraction. If the starship travels for 1 year in its coordinate system it will actually have traveled less than a year in the earth's coordinate system – possibly much less than a year in the earth's coordinate system – if the starship's speed has a β that is much greater than $\sqrt{2}$. *Thus people on the starship would appear to age more quickly than on earth.*

2.3.3 Combined Effect of Space and Time Dilation and Contraction

If we take account of the combined effects of space and time dilation and contraction respectively (eqs. 8.10 and 8.13) we find that

$$L/T = (\beta^2 - 1)L'/T' \qquad\qquad (2.54)$$

Now imagine a trip from earth to a star at the constant speed v of distance L and travel time T (neglecting initial acceleration and destination deceleration) and assume that v is such that β is greater than $\sqrt{2}$.[13] Then the effective speed as viewed from earth will be more than the effective speed as viewed in the starship. As a result the starship crew will believe it arrived at its destination more slowly than people on earth would think. So the combined effect of space and time dilation and contraction for β greater than $\sqrt{2}$ is to make the starship crew think it traveled more slowly to its destination. *If β is much greater than $\sqrt{2}$ then the starship crew will think it arrived much more slowly then the earth based observers – especially due the appearance of the square of β in eq. 2.54.*

2.4 Mass of a Starship – Tachyonic!

Lorentz transformations apply to the momentum of particles (and clumps of particles) as well as to the coordinates of particles. So we can take the simple case considered earlier of eq. 2.1 and relate the

---

[13] The effects of β greater than $\sqrt{2}$ were first noted in Blaha (2007a).

momentum of the starship in its (primed) coordinate system to the momentum of the starship as seen from earth (the unprimed coordinate system) using the same transformation as in eq. 2.1.

$$E' = \gamma(E - \beta c p_x) \tag{2.55}$$
$$p'_x = \gamma(p_x - \beta E/c)$$
$$p'_y = p_y$$
$$p'_z = p_z$$

where $\beta = v/c$, c is the speed of light, and $\gamma = (1 - \beta^2)^{-\frac{1}{2}}$. E and E' are the energies in the respective coordinate systems. And the spatial momenta are (p_x, p_y, p_z) and (p'_x, p'_y, p'_z) in the coordinate systems.

If we now assume β is greater than one then we obtain the superluminal transformation of momenta corresponding to eq. 2.2:

$$E' = \gamma_s(E - \beta c p_x) \tag{2.56}$$
$$p'_x = \gamma_s(p_x - \beta E/c)$$
$$p'_y = p_y$$
$$p'_z = p_z$$

where

$$\gamma_s = i\gamma = (\beta^2 - 1)^{-\frac{1}{2}} \tag{2.57}$$

In the example we are considering $p_y = p_z = 0$ and in the primed frame where the starship is at rest $p'_x = p'_y = p'_z = 0$ and $E' = M'c^2$ the mass of the starship in its rest frame coordinate system – the primed frame.

The momenta of a particle in a coordinate system are defined to have the form:

$$p^0 = E/c \qquad \text{and} \qquad p^i = m v^i \gamma/c \tag{2.58}$$

where i = 1, 2, 3 corresponding to the x, y, and z components of the velocity of the particle. The energy $E = \gamma mc^2$ and $\gamma = (1 - \beta^2)^{-\frac{1}{2}}$ with m the mass of the particle in the coordinate system.

Therefore the 4-vector inner product of the momenta (corresponding to eq. 2.2 for the coordinates) of a starship in its own rest frame is

$$p'^{0\,2} - \Sigma\, p'^{i\,2} = (E'/c)^2 = M'^2 c^2 \qquad (2.59)$$

where M' is the mass of the starship in the primed coordinate system.
Now E and p_x have the form given in eq. 2.58 with mass M in the
unprimed coordinate system. Substituting for E' using the transformation
law eq. 2.55 and

$$E = Mc^2\gamma \qquad\qquad p_x = Mv\gamma \qquad (2.60)$$

yields

$$(E'/c)^2 = [\gamma_s(E - \beta c p_x)/c]^2 = M^2 c^2 \gamma_s^2 \gamma^2 (1 - \beta^2)^2 = -\,M^2 c^2$$

and so

$$M^2 = -M'^2 \qquad (2.61)$$

*Thus a particle, and by extension an entire starship, is tachyonic from
the viewpoint of the earth's coordinate system. Tachyonic Starships –
meaning faster-than-light starships.*

3. Starship Thrust and Interstellar Travel

3.1 Exceeding the Speed of Light

In standard sublight relativistic dynamics the speed of a massive object cannot exceed the speed of light if the force applied to the object is real. In this section we will consider the case of a *complex-valued* force applied to an object (complex thrust) that causes the object to attain a complex velocity whose real part can exceed the speed of light. Complex valued forces have not been experimentally found in Nature as yet. However the motion of a particle inside a Black Hole is tachyonic. Its motion is determined by the gravitational curvature of space with the Black Hole. The Equivalence Principle of General Relativity tells us that the particle's gravitationally based motion can be viewed as a force acting on the particle in an inertial reference frame. (We experience the Equivalence Principle in this fashion every day – the force of gravity is equivalent to the gravitational curvature of space-time.) Now the gravitational equivalent force on a tachyon particle in a Black Hole must be a complex valued force or the tachyon's motion would not be tachyonic. Thus Black Holes and consequently Nature, do have complex valued forces.

In Blaha (2010a) we show that neutrinos and down-type quarks are tachyons. If we can harness down-type quarks (d quarks in particular) to create rocket thrust then we will have a mechanism for complex-valued thrust that can power starships faster than the speed of light.

Since a complex-valued rocket thrust will generate a complex-valued velocity and movement in space, the physical interpretation of complex velocities and distances must be addressed. In the previous chapter we showed that a superluminal transformation maps points with real coordinate values in one coordinate system to points with complex coordinate values in the target coordinate system. We then showed that the complex-valued coordinate points in the target coordinate system could be "rotated" to real valued coordinates using $\Pi_L(\mathbf{v}/v)$ (eq. 2.34). Thus the combined superluminal transformation and $\Pi_L(\mathbf{v}/v)$ transformation $\Pi_L(\mathbf{v}/v)\Lambda(\mathbf{v}/v)$ maps real coordinates to real coordinates. Complex coordinates are then merely an artifact of superluminal transformations.

However when we consider the path of a rocket with complex thrust that starts from a spatial point with real coordinates and, as it accelerates, traverses complex-valued spatial points a new issue arises: What is the physical meaning of these complex-valued spatial coordinates. Unlike the previous case of superluminal transformations one cannot simply use a global transformation to change the complex-valued points to points with real coordinate values. This is particularly clear if one considers a three point configuration: the earth, the rocket and the destination star. The earth and star have real valued coordinates in the earth coordinate system. The rocket in transit has complex coordinates at each point of its journey. In general, there is no global transformation that will make the coordinates of all three points real-valued. Therefore we conclude that complex coordinates are physically meaningful in this type of situation where one "point" is moving with a complex velocity. On this basis we will assume that space is three-dimensional with complex coordinate values in general.

Why haven't the complex values of coordinates been noticed before? Because objects with complex velocities have not been created and/or seen. To give an object a complex velocity we need either a highly curved space-time region (such as a Black Hole) with an event horizon that encloses the object so that we can't see it; or the movement of an object by a complex-valued force or thrust that would make the object traverse complex spatial points.

The second possibility can only be achieved with tachyon thrust or force. The only tachyons appear to be neutrinos or down-type quarks.[14] Neutrinos only interact via the Weak interaction and have strictly real momenta. So they are not capable of generating a complex thrust. Down-type quarks are confined within protons and neutrons. To create regions containing down-type quarks we would need collisions at enormous energies. We are just entering the experimental stage where this possibility can be realized. RHIC at Brookhaven National

---

[14] Blaha (2010a).

Laboratory and LHC at CERN have started creating quark-gluon fluids by colliding heavy ions such as gold and lead ions. Evidence for tachyonic quarks within the collision regions will hopefully be forthcoming soon.

Then a superluminal, tachyon drive starship with complex thrust becomes possible.

3.2 Superluminal Starship Dynamics

In this section we will consider a constant, propulsive force in a starship's rest frame that drives the starship from a sublight velocity to a superluminal velocity. The key factor in achieving a superluminal speed is evading the singularity in γ at $v/c = 1$. We accomplish this goal by having a complex force – a force with a real and imaginary part – that generates a complex acceleration, and thus a complex velocity, that "goes around" the singularity in the complex velocity plane.

We assume a constant, complex force exists in the rest frame of the starship due to the starship's thrust in the direction of the positive x' (and x) axis. The starship (primed coordinates) and earth (unprimed coordinates) coordinates have parallel axes as in Fig. 2.1. The spatial force in the positive x direction is

$$\mathbf{F'} = g\hat{\mathbf{x}} \tag{3.1}$$

where g is assumed to now be a complex constant.

The fourth component of the force (since force is a Lorentz 4-vector) is zero in the rocket's rest frame:

$$F'^{0} = 0 \tag{3.2}$$

Applying the inverse of the Lorentz transformation eq. 2.1 we find the force in the earth rest frame is

$$
\begin{aligned}
F^0 &= \gamma(F'^0 + \beta F'^x/c) = \gamma\beta F'^x/c = \gamma vg/c^2 \\
F^x &= \gamma(F'^x + \beta c F'^0) = \gamma F'^x = \gamma g \\
F^y &= F^z = 0
\end{aligned}
\tag{3.3}
$$

where $\beta = v/c$, c is the speed of light, and $\gamma = (1 - \beta^2)^{-\frac{1}{2}}$ as before. We again use the superscripts x, y, and z to identify the components of the spatial force. The spatial momentum of an object of mass m is

$$\mathbf{p} = \gamma m \mathbf{v} \qquad (3.4)$$

and the dynamical equation of motion is

$$d\mathbf{p}/dt = \mathbf{F} \qquad (3.5)$$

in the "earth" coordinate system resulting in

$$dp^x/dt = \gamma g \qquad (3.6)$$

with[15]

$$dp^y/dt = dp^z/dt = 0 \qquad (3.7)$$

The differential equation resulting from eq. 3.5 is

$$d(\gamma v)/dt = \gamma g/m \qquad (3.8)$$

which has the solution[16]

$$v = c\{1 - 2/(1 + ((c + v_0)/(c - v_0))\exp[2g(t - t_0)/(mc)])\} \qquad (3.9)$$

where the velocity is v_0 at time t_0. The complexity of the force constant g enables the velocity to exceed the speed of light.

   Before doing that we note that eq. 3.9 can easily be integrated to give the distance traveled in the x direction.

$$x = x_0 + (mc^2/g)\ln((1 - v_0/c + (1 + v_0/c)\exp[2g(t - t_0)/(mc)])/2) - c(t - t_0) \qquad (3.10)$$

or

---

[15] There is thrust in the y and z direction as well. To avoid getting distracted by the details of an exact calculation we approximate the force in those directions as zero.

[16] The velocity is entirely in the x-direction in this calculation. It can, and does, have complex values in this example. See footnote 45 in the discussion of our starship engine to see how the complexity of the value arises.

$$x = x_0 + (mc^2/g)\ln[(1 - v_0/c)/(1 - v/c)] - c(t - t_0) \qquad (3.11)$$

The complexity of g and thus the velocity causes x to be complex. The starship is then at a point x in complex space.

As a result superluminal travel to a distant star (or galaxy eventually) requires three phases in general. In the first phase the starship accelerates with a value for the thrust g that enables it to reach a high complex velocity whose real part was much greater than the speed of light. In the second phase the starship coasts to a point "not far" from the destination. At this point the starship is located in complex space. In the third phase the starship engines are turned on and the thrust set at a value that will bring the starship to its destination (located at a real valued coordinate position.) See Figs. 3.1 – 3.3.

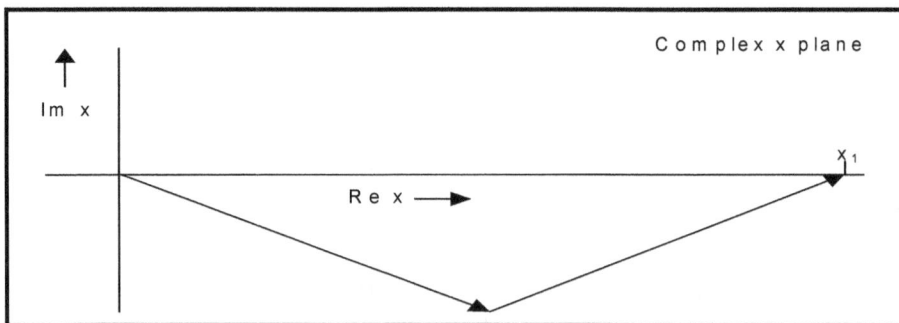

Figure 3.1. Rough depiction of the travel of a starship in the complex x dimension. The starship starts out in real space at x = 0. While accelerating, the starship is at a complex distance from its origin. After reaching cruising speed it turns off its superluminal engines until near its destination x_1. When nearing its destination it turns the superluminal engines back on, which brings it to its destination at the real distance x_1 at zero imaginary velocity and small real velocity.

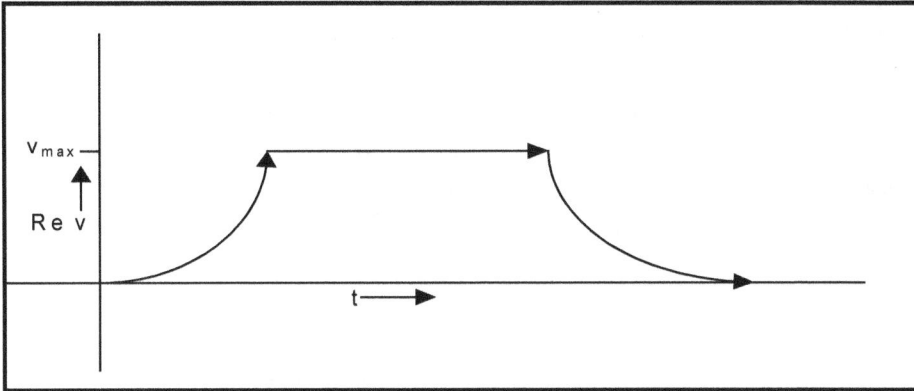

Figure 3.2. Depiction of the <u>real</u> part of a starship speed. There is an acceleration part to a desired maximum speed. Then the starship cruises at that speed until it reaches the vicinity of the destination. Then the starship drive decelerates it to a speed near zero so the starship can enter orbit around a star or planet.

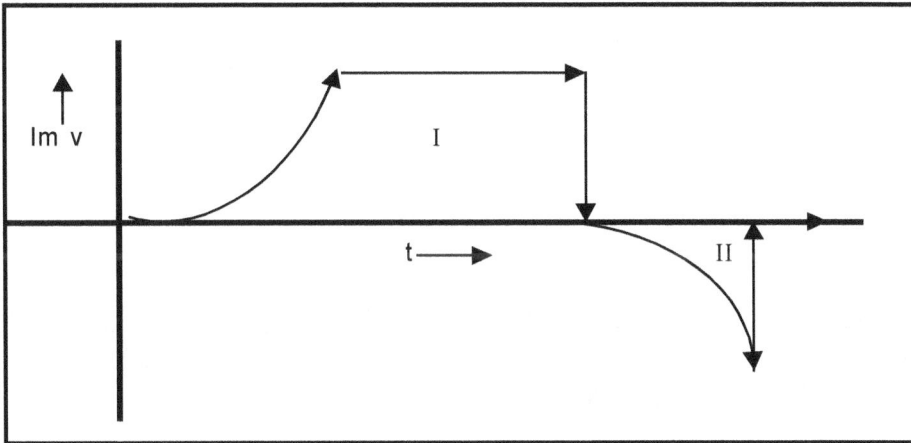

Figure 3.3. Rough depiction of a possible <u>imaginary</u> part of a starship speed. There is an acceleration part to high speed. Then the starship cruises at that speed until it reaches the vicinity of the destination. Then the starship drive decelerates causing the imaginary speed to decrease to zero so the starship has zero imaginary speed at the end of the deceleration. The net imaginary distance traveled is also forced to zero so that the starship ends in

real space. The areas of regions I and II are of equal magnitude to bring the imaginary part of the starship's location to zero.

3.3 Achieving High Superluminal Starship Velocities

To achieve the type of motion depicted in Figs. 3.1 – 3.3 the constant force value g must satisfy a special set of conditions. These conditions emerge from a consideration of the denominator of eq. 3.9:

$$1 + ((c + v_0)/(c - v_0))\exp[2g(t - t_0)/(mc)] \qquad (3.12)$$

If this denominator approaches zero then the speed v becomes infinite if g has an appropriate complex value. Let

$$g = g_1 + ig_2 \qquad (3.13)$$

If we wish the velocity to get very large (approach infinity) after some acceleration time interval $\triangle t = t - t_0$ we set

$$1 + ((c + v_0)/(c - v_0))\exp[2g\triangle t/(mc)] = 0 \qquad (3.14)$$

with the result

$$g_2 = n\pi mc/(2\triangle t) \geq 0 \qquad (3.15)$$

and

$$g_1 = (g_2/n\pi)\ln[(c - v_0)/(c + v_0)] \qquad (3.16)$$
$$= (mc/(2\triangle t))\ln[(c - v_0)/(c + v_0)] \leq 0$$

for n an odd, positive integer. Eqs. 3.15 and 3.16 enable the real part of the velocity to become infinite in the time interval $\triangle t$. We assume n = 1 in the following discussions. Substituting in eq. 3.9 we obtain

$$v = c\{1 - 2/(1 + [(c + v_0)/(c - v_0)]^{1 - (t - t_0)/\triangle t} e^{in\pi(t - t_0)/\triangle t}])\} \qquad (3.17)$$

3.4 Example of a Starship Accelerating from Sublight to Superluminal Speed

We will now consider a specific example of the acceleration phase of a starship to get a feeling for an *optimal* case.[17] First we note that the acceleration of a rocket of mass m with a propellant exhaust speed v_e in the rocket's rest frame is given by

$$dv/dt = (v_e/m) \, dm/dt \qquad (3.18)$$

and thus the constant g of eq. 3.1 is

$$g = mdv/dt = v_e \, dm/dt \qquad (3.19)$$

Since we intend to generate the thrust with a quark-gluon plasma producing an extremely high-energy exhaust we will choose the modest value of

$$dm/dt = 1 \text{ gm/sec} \qquad (3.20)$$

in our example. In one year of travel the amount of mass expelled is approximately 32 metric tons.[18] If we envision a starship of 10,000 metric tons[19] which we wish to accelerate to approximately 5,000c starting from a speed of .0001c (= 18.6 miles per second) within 30 days then the required value of g is approximately[20]

$$g = -0.3858c + i6060.166c \text{ gm-cm/sec}^2 \qquad (3.21)$$

---

[17] The value of the imaginary speed is quite high. In realistic cases such as those of the first generation of starships the value of the imaginary speed will undoubtedly be much lower – probably of the order of magnitude of the speed of light. Thus first generation starships will travel much more slowly but still much faster than the speed of light.

[18] The mass of the nuclear (or fusion) fuel needed to generate electricity to power the rings of the starship is another factor but will not be very large compared to the starship's mass.

[19] About one-fifth the mass of the ship Queen Elizabeth.

[20] The value of v_e (eq. 10.31) can be decreased by increasing dm/dt. For example if dm/dt = 100 gm/sec then v_e can be decreased by a factor of 100 with the same resulting starship motion.

where $c = 3 \times 10^{10}$ is the numerical value of the speed of light in cgs units.[21] The consequent exhaust speed is

$$v_e = -0.3858c + i6060.166c \qquad (3.22)$$

would be feasible using tachyonic quark acceleration to produce the thrust. We note there is no Special Relativistic limit on imaginary speed. The required exhaust speed can be reduced if the acceleration time is lengthened.

Thus the parameters of this starship example in cgs units are

m	10,000,000,000
c	29,979,245,800
g_2	181,679,202,956,833.
g_1	−11,566,067,091.
2g/mc (sec^{-1})	$-7.71604940843838 \times 10^{-11} + 1.21203317901235 \times 10^{-6}i$
2g/mc (day^{-1})	$-6.66666668889076 \times 10^{-6} + 0.1047196666666671$
v_0	0.0001c

The initial starship speed $v_0 = 0.0001c$ or 18.6 miles per second (roughly three times earth's escape velocity) is well within the capabilities of nuclear rockets.[22]

The real part of the velocity resulting from the above choice of parameters is shown in Fig. 3.4. The velocity increases slowly until near the 30 day point (earth time) where singular behavior occurs. The below table (and Fig. 3.4) show the velocity accelerating from $v_0 = 0.0001c$ at the beginning to its value 29.00006 days later is rather slow compared to the increase in velocity near the 30 day point.

Example of a Starship's Acceleration

time (days)	0	29.00006	29.00009	29.999992	29.9999998	29.99999996
Re v	.0001c	30.396c	121.59c	607.9c	5,066c	30,396c
Im v	0	477,465c	1,909,861c	9,549,305c	79,577,538c	$4.7745 \times 10^8 c$

---

[21] The mass ejected to generate the thrust to obtain a specific speed will be considered in section 7.6.

[22] Our starship will have superluminal engines for interstellar travel and nuclear rocket engines for interplanetary travel and maneuvering.

Note the real speed of 5066c at 29.9999998 earth days[23] (0.017 seconds short of 30 earth days) and of 30,396c at 29.99999996 earth days (0.0035 seconds short of 30 earth days). Having achieved the desired real speed the starship's engines turn off and the starship travels at constant speed until near the destination.

At 5066c any of the 100 or so known stars within 21 light years can be reached in about 1.5 days of coasting. There is also time needed to decelerate the starship so the actual travel time would be longer. At 30,396c any point in the galaxy could be reached in about 3 years. *Thus Milky Way travel times become comparable to 16ᵗʰ century oceanic travel times via ships to various parts of the world!*

3.5 The Velocity Near the Singularity

In calculating the speeds at various times close to the singularity[24] at the 30 day point we have used an approximation to eq. 3.17. Letting $t = t_1 + \tau$ where τ is small, and letting $\Delta t = t_1 - t_0$ then eq. 3.17 becomes

$$
\begin{aligned}
v &= c\{1 - 2/(1 + ((c + v_0)/(c - v_0))\exp[2g(\Delta t + \tau)/(mc)])\} \\
&= c\{1 - 2/(1 - \exp[2g\tau/(mc)])\} \\
&\approx c\{1 - 2/(1 - (1 + 2g\tau/(mc)))\} \\
&\approx c\{1 + (mc^2/g)(1/\tau)\} \\
&\approx c\{1 + (g*mc^2/|g|^2)(1/\tau)\}
\end{aligned}
\tag{3.23}
$$

---

[23] If the earth days were transformed to time on the starship the starship time would be dilated by approximately a factor of v/c.

[24] The rapid increase in speed near the 30 day point takes place in "earth" time. On the starship the time contraction effect causes starship time to increase rapidly as pointed out in the discussion of acceleration in subsection 10.5.4.1. Thus the apparent increase in speed on the starship is much less rapid.

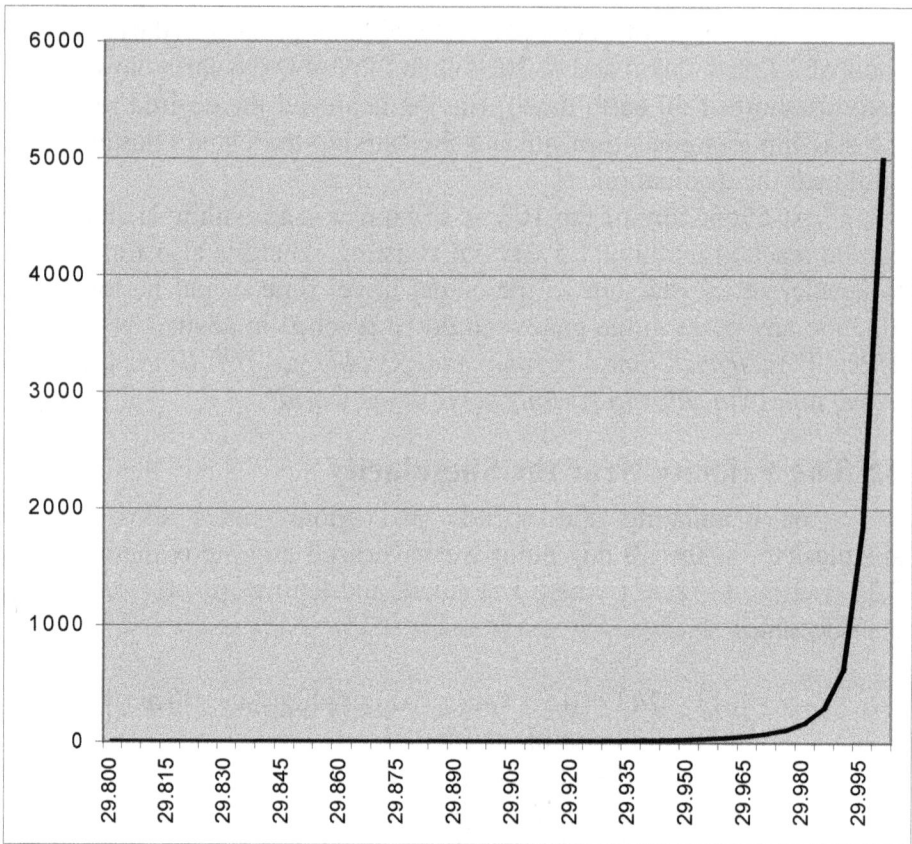

Figure 3.4. An example of the <u>real</u> part of the velocity of a starship on its 29th and 30th earth day of travel. The dynamics of this case are described in the text. The initial speed of the starship was 0.0001c. Its real speed reaches 5000c and beyond. Speed is measured in units of c; time is measured in earth days.

Given the signs of g_1 and g_2 (eqs. 3.15 and 3.16) we see that

- For small negative τ both the real and imaginary parts of v approach $+\infty$ as $\tau \to 0$.
- For small positive τ both the real and imaginary parts of v approach $-\infty$ as $\tau \to 0$ from above.

as displayed in Figs. 3.5 and 3.6. A starship can decide to switch off engines and "coast" at high speed towards the destination at some time close to the singularity point.

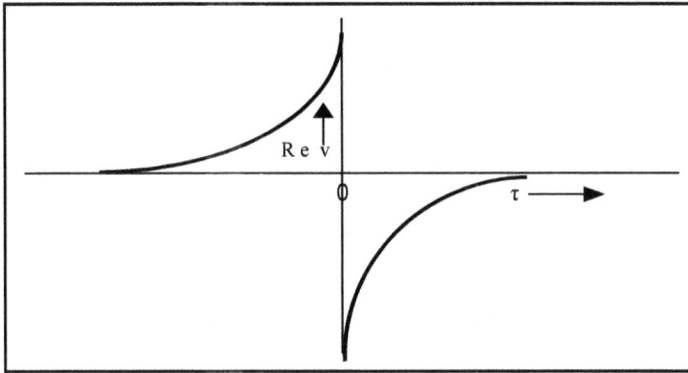

Figure 3.5. Qualitative plot of Re v from eq. 3.23 near the singularity at τ =0.

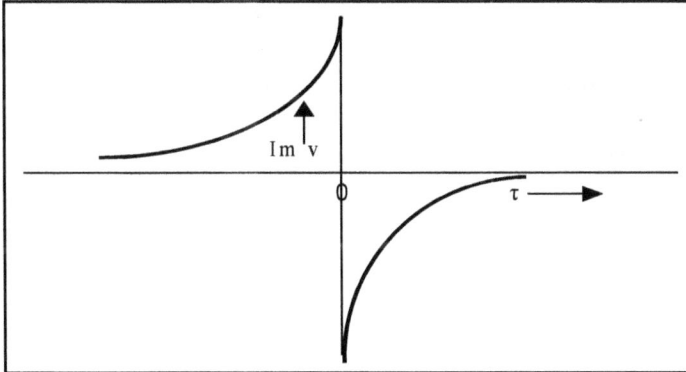

Figure 3.6. Qualitative plot of Im v from eq. 3.23 near the singularity at τ =0.

3.6 The Acceleration Experienced on the Starship

The rapid acceleration, particularly in the neighborhood of $\tau = 0$ raises the question of the inertial forces that would be experienced by passengers on the starship. *Remarkably the acceleration that the starship occupants feel is quite moderate due to the effects of relativity.*

The calculation of the maximum acceleration begins with the inverse of the relativistic transformation from earth coordinates to starship coordinates (eq. 3.3):

$$F'^0 = \gamma(F^0 - \beta F^x/c)$$
$$F'^x = \gamma(F^x - \beta c F^0) \qquad (3.24)$$
$$F'^y = F'^z = 0$$

which implies the acceleration felt by occupants in the starship is

$$a' = F'^x/m = \gamma(a - v\gamma/c^2) \qquad (3.25)$$

where a (in the earth coordinate frame) is given by the derivative of eq. 3.17

$$a = dv/dt$$
$$= -4(g/m)((c + v_0)/(c - v_0))\exp[2g(t - t_0)/(mc)]/\{1 + ((c + v_0)/(c -$$
$$- \ v_0))\exp[2g(t - t_0)/(mc)]\}^2$$
$$\qquad (3.26)$$

$$\simeq 2c/\tau + 4(g/m)$$

At $\tau = 29.9999998$ days we find using the approach of eq. 3.23[25]

$$a = -2{,}891{,}516{,}763{,}121 + 72{,}672i$$

Thus we find

$$a/g_E = -2950527309 + 74i$$

in ratio with the earth's gravitational acceleration g_E of about 980 cm/sec^2. The large value of the real acceleration <u>in earth coordinates</u> might lead a reader to question the ability of starship occupants to survive. However if we transform the earth coordinates acceleration to the equivalent starship coordinates acceleration we find the acceleration felt on the starship

---

[25] Kindly note that round off errors might lead the reader to slightly different results.

$$a' = a(5066^2c^2 - 1)^{-\frac{1}{2}} \approx a/(5066c) \qquad (3.27)$$
$$= (-1.94\times10^{-5} + i4.88\times10^{-13})g_E$$

has small real and imaginary parts compared to earth's gravitational acceleration. (We use the real speed in the example at $\tau = 29.9999998$ days: 5066c approximately.) *Thus the starship occupants will be safe.*

We have developed starship dynamics to allow speeds far in excess of the speed of light for human occupants based on the ability to produce complex valued thrust and the ability to have complex velocities and space. We address these issues later in this chapter.

3.7 Travel Time Experienced on the Starship – Suspended Animation

Another issue is the travel time experienced by the starship occupants. By subsection 2.3.2 it will appear much, much longer than that measured on earth. For example if $v = 5,000c$ then it will be approximately 5000 times longer. A 2 month trip from earth's view would take around 1,000 years from the view of the occupants of the starship. *Therefore a practical method of suspended animation must be found for long distance travel. A 4 month round trip to a star would require the starship occupants to be in suspended animation for approximately 2,000 years – starship time. With suspended animation they could be kept biologically roughly "in sync" with the earth measured travel time of 4 months (plus time spent at the destination) despite the starship elapsed time of 2,000+ years.*

3.8 Constant Superluminal Starship Travel

Having reached an enormous *real* speed such as a speed between 5000c and 30,000c we can turn off the superluminal engines. If we didn't, then the real speed would rapidly decline after passing "through" the singularity generated when the thrust is complex.[26]

The starship then moves at this constant speed in the absence of forces (and neglecting gravity and other minor perturbative forces). At a real speed of 5000c any place in the galaxy is a short travel time away.

---

[26] Due to the size of the starship, and other factors, the starship's velocity will not ever be precisely that of the singularity. Thus it is not an issue.

And nearby galaxies are reachable as well. Figure 3.7 shows the time required to reach various interesting destinations at 30,000c.

Destination	Distance (ly)	Approximate Travel Time (years)
To the other end of the Milky Way Galaxy	100,000	3
To the Center of the Milky Way	30,000	1
Large Magellenic Galaxy	150,000	5
Small Magellenic Galaxy	200,000	7
Andromeda Galaxy	2,000,000	70

Figure 3.7. "Coasting" part of travel time to various destinations at a real velocity of 30,000c.

Since much, much higher "coasting" velocities are also possible, almost the entire visible universe becomes accessible to Mankind. Mankind has an incredible future if it has the will to seize it.

Starship travel will be in complex space. That could be an advantage. In real space, with which we are familiar, there are asteroids, planets, stars, nebulae, black holes, and dust. If we travel in real space in a straight line we will have a significant probability of colliding with some of these objects. We certainly would collide with interstellar and intergalactic dust. The dust alone would severely damage a starship traveling at high speed. Over long distances it would also reduce the speed of the starship significantly.

In complex space it is not known if there are any objects similar to those found in our real space. If it is empty then the starship will avoid all the pitfalls of travel in real space. Thus the starship speed will not be impeded and it will not be damaged or destroyed by collisions with matter. We can hope that this is the case.[27]

---

[27] This hope may not be realized. As pointed out in Blaha (2004) p. 78-9 matter might exist in complex space. The "Great Attractor", a seemingly empty region of space has recently been shown to be drawing an enormous number of galaxies towards it. The Great Attractor may be an extraordinary large mass located in complex space – not the real space of our experience. As pointed out in Blaha (2008), and his earlier books, it appears that the universe began as a complex space that, due to symmetry breaking, broke into disjoint real and complex parts with no interaction between them except gravity, and except in localized regions such as inside hadrons in which quarks and gluons enjoy a small region of complex space.

3.9 Deceleration a Tachyonic Starship to Sublight Speeds

Eventually all journeys end so we will now examine the deceleration of a starship as it approaches its destination. We turn on the superluminal engine. The thrust is reversed ($g \rightarrow -g$) to decelerate. Otherwise the parameters of subsection 10.5.3 are in effect except v_0 which is now that of the 29.9999998 day point. We assume the starship was not slowed down by dust etc. and so traveled at constant velocity until the point when the superluminal engines were turned on for deceleration. The results are in the following table.

Example of Starship Deceleration

Time (days)	0	5	10	15	20	25	30
Re v		5067c	0.000249c	0.0001.33c	0.0001c	0.0000899c	0.0000893c	0.0001c
Im v	79,577,538c		3.73c	1.73c	1.00c	0.577c	0.268c	0.000001c

Note: Im v at the 30 day point in the above table appears non-zero due to round off error. Im v is, in fact, exactly zero at the 30 day point. Note also that the speed is now real after deceleration, and is the original speed it had prior to acceleration.

3.10 Imaginary Space Travel Distance

The imaginary distance traveled during the trip is also an issue. The imaginary distance traveled must exactly cancel by the end of the trip so that the starship ends the journey in purely real space. Since the bulk of the distance traveled occurs during the "coasting" phase a starship cannot truly coast during the entire coasting phase. It must generate a thrust during the last part of the coasting phase with a value for g that brings the imaginary distance traveled to zero at the destination. During this interval the complex number g will have a negative value for g_2 with g_1 equal to zero.

From eqs. 3.15 and 3.16, and assuming a purely imaginary thrust we see g' must be

$$g' = ig_2 = in\pi mc/(2\Delta t) \qquad (3.28)$$

where n is a positive, odd integer. We can decelerate *after* nearing the real distance of the destination. To that end we need a large negative imaginary velocity that will undo the positive imaginary distance traveled during the "coasting" phase. And we must "coast" an imaginary distance to bring the cumulative imaginary distance to zero at the destination. The velocity equation (eq. 3.17) gives a large negative imaginary velocity, which can be seen by using a variation on the approximation of eq. 3.23:

$$v = c\{1 - 2/(1 + ((c + v_c)/(c - v_0))\exp[2g'(\Delta t + \tau)/(mc)])\}$$
$$\simeq c\{1 - 2/[(2c/(c - v_c)) + 2ig'\tau/(mc)]\} \tag{3.29}$$

to order τ (for small τ) where v_c is the velocity during the coasting phase. Then since $|v_c| \gg c$ we can further approximate v to

$$v \simeq imc/(2g_2\tau) \tag{3.30}$$

which results in a large negative imaginary speed as τ approaches zero (the singularity point) from below.

If we now let the starship coast to a zero value for imaginary distance and follow the procedure of subsection 3.9 to obtain a zero value for imaginary speed we achieve our goal of traveling a very large real distance in light years in a relatively short time.

3.11 Starship Travel Summation

The preceding example considered the case of a specific travel time of 30 days to near a singularity in v that enables us to achieve enormous superluminal speeds. We have also seen how to coast at these speeds and how to decelerate to small, sublight speeds. Other cases with different travel times to the singularity in v have similar qualitative features although the numerics are different. Planning a starship flight requires a complex program specifying a combination of all of these features of acceleration, coasting, and deceleration in such a way as to minimize energy and fuel usage.

Thus we have demonstrated that realistic superluminal travel is possible in principle if

1. Complex thrust engines are possible.
2. It is possible to travel in complex space.
3. Complex space does not have insurmountable obstacles to superluminal speeds such as interstellar dust or massive

objects that could lead to collisions or rapid erosion of the starship.

We will address item 1 in the following chapters of the book and refer readers to Blaha (2010a), (2008) and (2004) that suggest very strongly that complex space exists.

4. Overview of Quark Drive Starships

In the preceding chapter we saw the enormous possibilities of tachyonic starships to explore and colonize the galaxy. This chapter describes the basic concepts of the tachyonic, quark drive engines of a starship. The following chapters describe the design and its mathematics in more detail.

4.1 Overall Design of a Quark Drive Starship

The design of a starship is a difficult matter at best due to major scientific and engineering breakthroughs that are required. However one can envision an overall design assuming the necessary breakthroughs take place. The overall characteristics of a "simple" starship are described below. Fig. 4.1 shows a tentative view of the proposed starship.

1. A starship would have a spherule[28] collider assembly that would accelerate and collide spherules at high energy to produce "large" quark-gluon fireballs.

2. "Just before" collision the spherules would be compressed to nuclear density by an array of powerful lasers such as those used to compress pellets in fusion energy reactors. The result of the collision of the compressed spherules would be a macroscopic quark-gluon fireball whose "guided" expansion would furnish a high, complex-valued thrust.

3. Naturally it would be reasonable to have the colliding rings surround the outer edge of the starship to reduce the required

---

[28] A spherule is a small sphere. In the present case we anticipate charged U-238 spherules (with perhaps a mass of the order of 0.01 mg) that collide in a colliding ring assembly that has an overall similarity to ion colliding rings such as those at RHIC and LHC.

bending magnet strength requirements. Thus the starship shape displayed in Fig. 4.1.

4. The interior region of the starship would hold fuel, life support equipment and cargo, nuclear/fission power generators, living quarters, cargo areas, and (perhaps four) nuclear engines for "short distance" travel within star systems. the starship. The decision will be based on maximizing the efficiency and reliability of the nuclear rocket configuration.

5. The starship is not intended to land on large planets. It would contain shuttles for that purpose.

Figure 4.1. A visualization of a starship. The outer disk contains the colliding hadron rings which generate quark-gluon fireballs in a "combustion chamber" that expands through a "rocket nozzle"

generating a complex-valued thrust that enables the starship to exceed the speed of light. The four nuclear engines depicted on the underside of the ship provide "intra-solar system" speeds.

6. The nuclear rockets would be used for "short" distance travel within solar systems. These rockets could be either ion drive rockets where ions were accelerated by electromagnetic forces generated by electricity provided by the ship's nuclear reactors. Or the rockets could simply use nuclear reactors to directly heat a fuel such as hydrogen whose ejection would provide thrust for the starship.

5. Superluminal Acceleration

In this chapter we will consider the acceleration of superluminal particles by electromagnetic fields.[29] Since superluminal particles such as quarks can have complex-valued velocities they have significantly different behavior under electromagnetic forces. This chapter is written with a view to the distant future when quark-gluon fireballs can be accelerated. Quarks, having charge, can be accelerated by electric and magnetic fields, and "pull" the electric charge zero gluon fields along with them. *The non-technical reader may skip this chapter.*

5.1 Lorentz Force Equations

The key equation for the motion of a charged particle under the influence of an electromagnetic field is the Lorentz force equations:

$$d\mathbf{p}/dt = qe(\mathbf{E} + \mathbf{v} \times \mathbf{B}/c) \tag{5.1}$$

$$dE/dt = qe\mathbf{v}{\cdot}\mathbf{E} \tag{5.2}$$

where **E** and **B** are the electric and magnetic fields, and qe is the charge of the particle.

An issue that arises is the requirement that the energy of a particle be real. If the particle velocity is complex as it is in general for quarks:[30]

---

[29] The reader may feel that many of the parts of this chapter are somewhat elementary but upon examination will se that the superluminal nature of the particles introduces subtle points of difference from the corresponding subluminal kinematics. **We note that the discussions of superluminal particles in this, and subsequent chapters, applies to quasi-free quarks within a quark gluon plasma or fluid such as is generated in high energy hadron collisions at RHIC or the LHC at present. The quarks are confined to a region that ranges from spherical to elongated. This region has complex spatial coordinates and momenta. The region will be modeled as a "bag" – that is a bubble-like region with a surface. The quark acceleration topics in this chapter implicitly take place within an extended bag.**

[30] Blaha (2010a) and earlier works.

$$\mathbf{v} = \mathbf{v}_r + i\mathbf{v}_i \qquad (5.3)$$

where

$$\mathbf{v}_r \cdot \mathbf{v}_i = 0 \qquad (5.4)$$

then for the particle energy to be *real* as time evolves

$$\mathbf{v}_i \cdot \mathbf{E} = 0 \qquad (5.5)$$

must hold by eq. 5.2 since the electric field $\mathbf{E}$ is real. If $\mathbf{v}_i \cdot \mathbf{E} \neq 0$ then the energy becomes complex as time progresses. Then the interpretation of a complex energy becomes an issue. The issue is resolved by noting that a superluminal transformation at time t to a coordinate system moving at a velocity of $\mathbf{v}_i(t)$ makes the imaginary velocity in the new coordinate system zero so that dE/dt is instantaneously real. Application of $\Pi'_L = \text{diag}(e^{i\chi}, 1, 1, 1)$ of eq. 2.38 to the transformed particle 4-momentum with a suitable choice of χ will make the energy E(t) a real number. Thus the physical interpretation of a complex particle energy is clear when the coordinate system is boosted to a coordinate system by a superluminal transformation followed by an appropriate choice of Π'_L.

5.2 Superluminal Particle Acceleration under a Constant Electric Force

We will consider the case of a superluminal particle moving in a constant electric field where the real part of the particle momentum is parallel to the electric field (and thus the imaginary part of the velocity is perpendicular to the electric field).[31] Since the electric force is real, the Lorentz force equation eq. 5.1 becomes

$$\begin{aligned} d\mathbf{p}_r/dt &= qe\mathbf{E} \\ d\mathbf{p}_i/dt &= 0 \end{aligned} \qquad (5.6)$$

where $\mathbf{p}_r$ is the real part and $\mathbf{p}_i$ the imaginary part, of the 3-momentum. Eq. 5.6 implies the imaginary part of the particle 3-momentum is

---

[31] The change in the imaginary velocity with time is due to eqs. 5.6 through 5.10 that link the time dependence of $\mathbf{v}_r$ and $\mathbf{v}_i$.

constant in time. However the imaginary velocity changes in magnitude and direction with time. [32] The real part

$$\mathbf{p}_r = \gamma m \mathbf{v}_r \qquad (5.7)$$

when inserted in eq. 5.6 gives the dynamical equation

$$md(\gamma \mathbf{v}_r)/dt = qe\mathbf{E} \qquad (5.8)$$

where

$$\gamma = (1 - \beta^2)^{-\frac{1}{2}} \qquad (5.9)$$

with

$$\beta = |\mathbf{v}|/c = (\mathbf{v} \cdot \mathbf{v})^{\frac{1}{2}}/c = (\mathbf{v}_r \cdot \mathbf{v}_r - \mathbf{v}_i \cdot \mathbf{v}_i)^{\frac{1}{2}}/c \qquad (5.10)$$

Eq. 5.8 is easily integrated to give

$$\mathbf{p}_r(t) = \mathbf{p}_{r0} + qe\mathbf{E}(t - t_0) = \gamma m \mathbf{v}_r(t) \qquad (5.11a)$$

where $\mathbf{p}_{r0}$ is the momentum at time t_0. Note the imaginary part is constant and equal to its initial value in this case

$$\mathbf{p}_i(t) = \mathbf{p}_{i0} = \gamma m \mathbf{v}_i(t) \qquad (5.11b)$$

The change in energy with time can be obtained from eq. 5.2:

$$E(t) = E_0 + qe\mathbf{E} \cdot (\mathbf{r}(t) - \mathbf{r}_0) \qquad (5.12)$$

where E_0 is the energy at time t_0, $\mathbf{r}(t)$ is the position at time t, and $\mathbf{r}_0$ is the position at time t_0:

$$\mathbf{r}(t) = \int_{t_0}^{t} dt \, \mathbf{v}(t) \qquad (5.13)$$

Note $\mathbf{r}(t)$ is a complex 3-vector in general since the velocity is a complex 3-vector. Eq. 5.12 is identical in form to the energy calculated for "normal" particles, the change in particle energy equals the change in potential energy, but differs in that $E(t)$ is complex in general in the

---

[32] Note γ depends on time due to its dependence on the instantaneous velocity; eqs. 5.10 and 5.11.

present case since the coordinate difference $\mathbf{r}(t) - \mathbf{r}_0$ is complex in general.

We conclude this subsection by noting again that the imaginary part of a particle's spatial momentum is changed by an electric field in general. We can calculate the real and imaginary velocities from eqs. 5.11. Defining

$$\alpha(t) = \gamma|\mathbf{v}_r(t)| = |\mathbf{p}_{r0} + qe\mathbf{E}(t - t_0)|/m \qquad (5.14)$$

$$\delta = \gamma|\mathbf{v}_i(t)| = |\mathbf{p}_{i0}|/m \qquad (5.15)$$

we find after some algebra

$$|\mathbf{v}_r(t)| = \alpha/[1 + (\alpha^2 - \delta^2)/c^2]^{\frac{1}{2}} \qquad (5.16)$$

$$|\mathbf{v}_i(t)| = \delta/[1 + (\alpha^2 - \delta^2)/c^2]^{\frac{1}{2}} \qquad (5.17)$$

for $c^2 + (\alpha^2 - \delta^2) > 0$. Although the imaginary momentum is constant, the imaginary velocity changes magnitude, but not its direction perpendicular to $\mathbf{E}$ in this case, due to its appearance in γ. For sublight motion, as $t \to \infty$, $\alpha(t) \to \infty$ and $|\mathbf{v}_i(t)| \to 0$ while $|\mathbf{v}_r(t)| \to c$ in conformity with the Special Theory of Relativity limiting particle speeds to less than the speed of light. However if $c^2 + (\alpha^2 - \delta^2) < 0$ then as t approaches a value such that $c^2 + (\alpha^2 - \delta^2) \to 0$, then $|\mathbf{v}_r(t)| \to -i\infty$ and $|\mathbf{v}_i(t)| \to -i\infty$. In this region both $|\mathbf{v}_r(t)|$ and $|\mathbf{v}_i(t)|$ are imaginary since they correspond to faster than light motion. The Π'_L transformation must be applied to these values to obtain the physical values of the real and imaginary velocities.

The transition from sublight to superluminal particle motion is best understood from the expression for γ:

$$\gamma = [1 + (\alpha^2 - \delta^2)/c^2]^{\frac{1}{2}} \qquad (5.18)$$

following from eqs. 5.9 and 5.10. Note that α increases with time t while δ is constant. We now consider the sublight case and then the superluminal case.

Sublight Motion ($|\mathbf{v}_r(t)| < c$)

In this case, supposing the motion begins at a sublight speed, γ begins as a small real quantity (assuming δ is sufficiently small) and increases with time. If $|\mathbf{v}_i(t)| = 0$, then as $|\mathbf{v}_r(t)| \to c$ we see $\gamma|\mathbf{v}_r(t)| = \alpha \to \infty$ and $\gamma \to \infty$. If $|\mathbf{v}_i(t)| \neq 0$, then as $|\mathbf{v}_r(t)| \to c$ we see $\gamma|\mathbf{v}_r(t)| = \alpha$ is finite as is γ. However from eqs. 5.9 and 5.10 we see that $\gamma \to \infty$ and $\alpha \to \infty$ at

$$|\mathbf{v}_r(t)| = [1 + |\mathbf{v}_i|^2/c^2]^{\frac{1}{2}} \qquad (5.19)$$

Thus the real part of the velocity can exceed the speed of light. The real part of the speed makes γ singular when eq. 5.19 is satisfied. The real part of the velocity is limited by the value of the imaginary part of the velocity. If the imaginary part of the velocity is increasing then the limit on the real part is correspondingly increasing.

Superluminal motion

If

$$|\mathbf{v}_r(t)| > [1 + |\mathbf{v}_i|^2/c^2]^{\frac{1}{2}}$$

then γ has an imaginary value and consequently the real and imaginary parts of the particle's momentum are imaginary as well. As noted earlier the application of $\Pi'_L = \text{diag}(e^{i\chi}, 1, 1, 1)$ of eq. 2.38 to the particle 4-momentum with a suitable choice of χ will make the energy $E(t)$ and 3-momenta real physical numbers.

As time increases, $\alpha \to \infty$, $v \to \infty$, the energy $E \to 0$, and the momentum $|\mathbf{p}| \to m$. Thus increasing the velocity well beyond $[1 + |\mathbf{v}_i|^2/c^2]^{\frac{1}{2}}$ has a vastly different effect on a particle's energy and momentum then in the sublight case. The behavior of a particle under a constant electric force is analogous to the case of an accelerating starship that we discussed in chapter 3.

5.3 Superluminal Particle Acceleration under a Constant Magnetic Force

We will now consider a superluminal particle traveling through a constant magnetic field. The Lorentz force equation is

$$d\mathbf{p}/dt = qe\mathbf{v} \times \mathbf{B}/c \qquad (5.20)$$

if we decompose the velocity and momentum into components parallel and perpendicular to B:

$$\mathbf{v} = \mathbf{v}_r + i\mathbf{v}_i = \mathbf{v}_\perp + \mathbf{v}_\| = \mathbf{v}_{\perp r} + \mathbf{v}_{\| r} + i\mathbf{v}_{\perp i} + i\mathbf{v}_{\| i} \qquad (5.21)$$

$$\mathbf{p} = \mathbf{p}_r + i\mathbf{p}_i = \mathbf{p}_\perp + \mathbf{p}_\| = \mathbf{p}_{\perp r} + \mathbf{p}_{\| r} + i\mathbf{p}_{\perp i} + i\mathbf{p}_{\| i} \qquad (5.22)$$

then eq. 5.20 becomes

$$d\mathbf{p}_\|/dt = 0 \qquad (5.23)$$

$$d\mathbf{p}_\perp/dt = qe(\mathbf{v}_{\perp r} + i\mathbf{v}_{\perp i}) \times \mathbf{B}/c \qquad (5.24)$$

Eq. 5.22 implies $\mathbf{p}_\|(t)$ is a constant. Taking real and imaginary parts of eq. 5.24 yields

$$d\mathbf{p}_{\perp r}/dt = qe|\mathbf{v}_{\perp r}||\mathbf{B}|\mathbf{u}/c \qquad (5.25)$$

$$d\mathbf{p}_{\perp i}/dt = qe|\mathbf{v}_{\perp i}||\mathbf{B}|\mathbf{w}/c \qquad (5.26)$$

where $\mathbf{u}$ and $\mathbf{w}$ are unit vectors in the directions of $\mathbf{v}_{\perp r} \times \mathbf{B}$ and $\mathbf{v}_{\perp i} \times \mathbf{B}$ respectively. Thus in a manner similar to the purely real velocity case we find the superluminal particle is deflected both in its real and imaginary motion in a plane to which the magnetic field **B** is perpendicular.

Taking the inner product of eq. 5.25 with $\mathbf{p}_{\perp r}$ and eq. 5.26 with $\mathbf{p}_{\perp i}$ we see that

$$d(\mathbf{p}_{\perp r}\cdot\mathbf{p}_{\perp r})/dt = d(\mathbf{p}_{\perp i}\cdot\mathbf{p}_{\perp i})/dt = 0 \qquad (5.27)$$

so that the magnitudes of the real and imaginary perpendicular components of the velocity and momentum are constant, and remain equal to their initial values, in a constant magnetic field. Eq. 5.27 implies

$$d(\gamma\mathbf{v}_{\perp r})^2/dt = d(\gamma\mathbf{v}_{\perp i})^2/dt = 0 \qquad (5.28)$$

or

$$\gamma|\mathbf{v}_{\perp r}| = c_r$$

$$\gamma|\mathbf{v}_{\perp i}| = c_i \qquad (5.29)$$

$$|\mathbf{v}_{\perp r}|/|\mathbf{v}_{\perp i}| = v_r/v_i = c_r/c_i$$

where c_r and c_i are constants. As a result the magnitudes $v_{\perp r}$ and $v_{\perp i}$ are constants and equal to their initial values $v_{\perp r0}$ and $v_{\perp i0}$ respectively.

Next we note that eq. 5.4 implies

$$\mathbf{v}_{\|r}\cdot\mathbf{v}_{\|i} + \mathbf{v}_{\perp r}\cdot\mathbf{v}_{\perp i} = 0 \tag{5.30}$$

If $\mathbf{v}_{\|r} = 0$ (The initial real motion of the particle is perpendicular to the magnetic field **B**.) or $\mathbf{v}_{\|i} = 0$ then

$$\mathbf{v}_{\perp r}\cdot\mathbf{v}_{\perp i} = 0 \tag{5.31}$$

and the particle executes a circular motion in the plane perpendicular to **B** (assumed to point in the positive z direction.)

We can then parameterize the real and imaginary perpendicular components of $\mathbf{v}_\perp$ by

$$\mathbf{v}_{\perp r} = v_{\perp r}(\hat{x}\cos\theta_r + \hat{y}\sin\theta_r) \tag{5.32}$$
$$\mathbf{v}_{\perp i} = v_{\perp i}(-\hat{x}\sin\theta_i + \hat{y}\cos\theta_i) \tag{5.33}$$

where by eq. 5.31

$$\theta_i = \theta_r \pm \pi \tag{5.34}$$

resulting in

$$\mathbf{v}_{\perp i} = v_{\perp i}(\hat{x}\sin\theta_r - \hat{y}\cos\theta_r) \tag{5.35}$$

If we extract the $\hat{x}$ components of the real and imaginary parts then the real and imaginary parts of the velocity have the ratio

$$v_{\perp r}\cos\theta_r/(v_{\perp i}\sin\theta_r) = (c_r/c_i)\cot\theta_r \tag{5.36}$$

Thus we can change the ratio of the real and imaginary parts of a particle's velocity with a constant magnetic field. This could, of course, be useful in piloting a starship. For example, at $\theta_r = 0$ the real part of the perpendicular velocity in the $\hat{y}$ direction has the magnitude 0 and the imaginary part of the perpendicular velocity has the magnitude $v_{\perp i0}$. At $\theta_r = \pi/2$ the real part of the perpendicular velocity in the $\hat{y}$ direction has the magnitude 0 and the imaginary part of the perpendicular velocity has the magnitude $v_{\perp i}$. So if the particle was extracted in the $\hat{y}$ direction (or some other direction) one could adjust the real and imaginary parts of the particle velocity.

Since the magnitudes, $v_{\perp r}$ and $v_{\perp i}$, are constant in time eq. 5.25 becomes

$$\gamma d[(\hat{x} \cos\theta_r + \hat{y} \sin\theta_r)]/dt = qe(|\mathbf{B}|/c)[-\hat{y}\cos\theta_r + \hat{x} \sin\theta_r] \qquad (5.37)$$

yielding[33]

$$\theta_r = \theta_{r0} - qe[|\mathbf{B}|/(\gamma c)]t \qquad (5.38)$$

Also since $v_{\perp r}$ and $v_{\perp i}$, are constant in time the particle trajectory in the plane will be a spiral with a linear time dependence:

$$r_r = r_{r0} + v_{\perp r0}t \qquad (5.39)$$
$$r_i = r_{i0} + (c_i/c_r)v_{\perp r0}t \qquad (5.40)$$

where r_r and r_i the radial distances from the center of the coordinate system.

The radii of curvature of the real and imaginary motion can be shown to be

$$\rho_r = cp_\perp/(|q|e|\mathbf{B}|) \qquad (5.41)$$
$$\rho_i = cp_\perp/(|q|e|\mathbf{B}|) \qquad (5.42)$$

from eqs. 5.25 and 5.26.

5.4 Manipulating the Real and Imaginary Parts of a Superluminal Particle's Acceleration and Velocity: Gluon Forces

The discussion in the previous subsections show that an electric field can change the real and imaginary parts of a particle's velocity and a uniform, constant magnetic field causes rotation of the real and imaginary parts of a particle's motion. The combination of these fields can be used for acceleration and the adjustment of the ratio of the real to the imaginary velocities.

---

[33] As the particle velocity becomes extremely large θ_r changes more and more rapidly since $\gamma^{-1} \sim$ v/c as v → ∞.

6. Quark-Gluon Fluids and Accelerator Thrust

6.1 Relativistic Heavy Ion Collisions

The collisions of highly relativistic heavy ions such as lead-lead, gold-gold, and uranium-uranium collisions is a relatively new field in experimental particle/nuclear physics. Experiments at the SPS (CERN) collided gold and lead ions at an energy of 17 GeV per nucleon and experiments at RHIC (Brookhaven National Laboratory) collided lead and gold ions at 130 GeV per nucleon. At the time of this writing initial ion collision experiments have begun at LHC (CERN) using colliding lead ions[34] – each with an energy of 1380 GeV per nucleon. All of these experiments indicate something new and exciting is happening in highly relativistic heavy ion collisions that does not happen in p-p (proton-proton) collisions. Colliding ions temporarily create a fireball containing quarks and gluons in a macroscopic state known as a perfect fluid. Within this region the quark-gluon fluid acts as if it had little or no viscosity (friction).[35] After a short time[36] 5 – 7 fm/c, during which this perfect fluid is expanding, the fluid enters the "Freeze-Out" state dispersing into numerous particles that stream out of the collision region.

During the time that the quark-gluon perfect fluid exists it is in thermal equilibrium and the laws of thermodynamics apply as well as the equations of ideal relativistic hydrodynamics. Many papers[37] now exist in the literature that describe the nature and evolution of the macroscopic quark-gluon fluid as well as the freeze out state that follows. These papers as well as the experimental results that can be expected from LHC studies in the next few years will provide a sound basis for the

---

[34] About 400 nucleons collide with each other in lead-lead and gold-gold collisions.

[35] The particles in the quark-gluon fluid rapidly reach thermal equilibrium after an ion collision through an as yet undetermined mechanism.

[36] Time is measured in fm/c = 3.34×10^{-24} sec. The symbol fm represents a distance of one fermi = 10^{-15} m. c is the speed of light.

[37] See E. Shuryak, arXiv:0807.3033 (2007) and references therein.

understanding of the dynamics of the quark-gluon state and the subsequent freeze out state.

The details of these results is not of immediate interest for our goal of designing a starship engine. Rather we want to investigate an extension of the present experimental LHC configuration that can serve as a prototype for a starship drive. In brief we view the drive as composed of an intersecting storage ring(s) setup that collides ionized heavy elements (in small spheres – spherules) in a chamber defined by confining laser or particle beams that is similar in concept to a rocket combustion chamber and nozzle. The macroscopic fireball generated by the collision(s) contains a large number of "unconfined" quarks with complex velocities that are funneled by the beams into complex rocket thrust for the starship as described in chapters 3 and 4. The major details of a starship engine design appear in chapter 7.

A large amount of data has been at SPS and RHIC. For our purposes the data of interest can be summarized as:

1. The generated fireball is a macroscopic, hot, dense, strongly interacting perfect fluid containing unconfined quarks and gluons with complex spatial momentum.

2. Data[38] from RHIC include:

    Maximum diameter of fireball for lead at various energies – about 5 fm

    Initial size of fireball – a diameter of 0.8 fm for lead

    Transverse v/c at edge of fireball is about .5 for times τ: 3 – 9 fm/c

3. Some Estimates:[39]

    • Hydrodynamic relaxation time are below 1 fm/c

    • Expansion rate – "Hubble fireball constant" τ^{-1} = $\partial_\mu u^\mu$ where u^μ is the fluid 4-velocity. τ is estimated to be of the order of 1 to several fm/c

    • Fireball lifetime – Hubble fireball time until freeze out 5 – 7 fm/c

[38] K. Dusling and D. Teaney, arXiv; 0710.5932 (2007).
[39] P. F. Kolb and U. Heinz, arXiv:nucl-th/0305084 (2003)

- Equilibration time (time for hydrodynamic fireball stage to begin) 0.6 fm/c
- Initial fireball volume expansion is linear in t but grows to t^3
- Fireball energy density 25 GeV/fm^3 at fireball center at 130 GeV
- Ellipticity of a fireball typically is a 10% - 20% effect
- The transition to quark-gluon perfect fluid occurs about 6.5 times cold nuclear density at 0.14 nucleons/fm^3
- The energy density in the fireball center rises roughly linearly with GeV/nucleon: density (GeV/ fm^3) = 0.19×(GeV/nucleon)
- The time to cross a nucleus of radius R $\cong$ 1.2A$^{1/3}$ is approximately 6 fm/c for lead or uranium

4. LHC – ALICE Group Results[40]

The ALICE group lead-lead experiment[41] at LHC found that the elliptic flow of charged particles at the LHC energy of 2760 GeV/nucleon was 30% greater than that encountered at the RHIC where the energy was 130 GeV/nucleon. The quark-gluon fluid at LHC energies also displays little friction and thus is a perfect fluid. Future studies at the LHC are expected to yield extremely precise data on the nature and behavior of the quark-gluon fluids.

6.2 Spherule Accelerators

In the previous section we considered hadron-hadron collisions for the case of heavy ions such as lead, gold and uranium that consist entirely of a nucleus since each ion is stripped of all its electrons. The radius of a heavy ion nucleus such as lead or uranium is about 6 fm. We

---

[40] CERN Courier **51**, 7 (April, 2011).
[41] K. Aamodt et al, Phys. Rev. Lett. **105**, 252302 (2011).

now wish to consider spherule-spherule accelerators and collisions since we wish to maximize the mass expelled per second of the starship thrust.

The first issue is the magnet strength needed to confine spherules to a "circular" orbit realizing that there are many additional magnetic ensembles needed to stabilize the orbits of cycling spherules in a snchrocyclotron.[42] Another issue that we will not address is the mechanism to accelerate and inject spherules into the main synchrotron ring. (See Fig. 8.1 for a diagram of the LHC – Large Hadron Collider setup showing the many stages of acceleration from zero to the initial hadron energy in the main 27 km ring.)

6.2.1 Hadron Collider "Turning" Magnets

The magnitude $|\mathbf{B}|$ of the magnetic field required to maintain a radius of curvature ρ is given by

$$|\mathbf{B}| = pc/(|q|\rho) \qquad (6.1)$$

where p is the particle's momentum and q is the particle's charge. The ratio of the required magnetic field for a spherule $|\mathbf{B_s}|$ to the magnetic field $|\mathbf{B_h}|$ required for a hadron ion to have the same radius of curvature ρ is

$$|\mathbf{B_s}|/|\mathbf{B_h}| = (p_s/p_h)(|q_h|/|q_s|) \qquad (6.2)$$

Assuming the hadron and ion velocities are equal, then eq. 6.2 becomes

$$|\mathbf{B_s}|/|\mathbf{B_h}| = (m_s/m_h)(|q_h|/|q_s|) \qquad (6.3)$$

where m_s and m_h are the masses of the spherule and ion respectively. If the spherule has a radius of 0.01 mg and is composed of U-238 then its mass ratio with a single U-238 ion is the number n of uranium atoms in the spherule. The charge of a fully stripped U-238 ion is 92e. If one electron is stripped from each U-238 atom (on average) in a .01 mg spherule then the charge on the spherule is 0.34 μC (micro-Coulombs). The ratio of the required spherule accelerator magnetic field to the ion's magnetic field in this case is 92 times the ion's magnetic turning field

---

[42] See Lee (2004) and other accelerator books.

$$|\mathbf{B_s}|/|\mathbf{B_h}| = n(92/n) = 92 \qquad (6.4)$$

or, more generally, for an element of atomic number A

$$|\mathbf{B_s}|/|\mathbf{B_h}| = A \qquad (6.4a)$$

Thus the turning magnetic field strength for a spherule is the element's atomic number times that required for an ion of the same element. Magnets with a field strength about 100 times that of the LHC turning magnets are possible.

6.2.2 Electric Acceleration of Spherules

RF generators accelerate particles[43] in a synchrotron using intense electric fields. The process may be simply viewed as an acceleration of a particle with a change in energy E given by the change in potential energy in the rf generator in the amount

$$q\Delta V$$

where ΔV is the change in electric potential energy. The ratio of the energy gain per atom/ion of the spherule to an individual ion is

$$(q_s/q_h)/n = (n/92)/n = 1/92 \qquad (6.5)$$

or, more generally, for an element of atomic number A

$$(q_s/q_h)/n = 1/A \qquad (6.5a)$$

since we assume one electron is removed per atom in the spherule and each ion has 92 electrons stripped from it. Thus the energy gain per atom in the spherule is 1/92 that of an ion under the same potential difference.

---

[43] Both the real and imaginary parts of particle velocities.

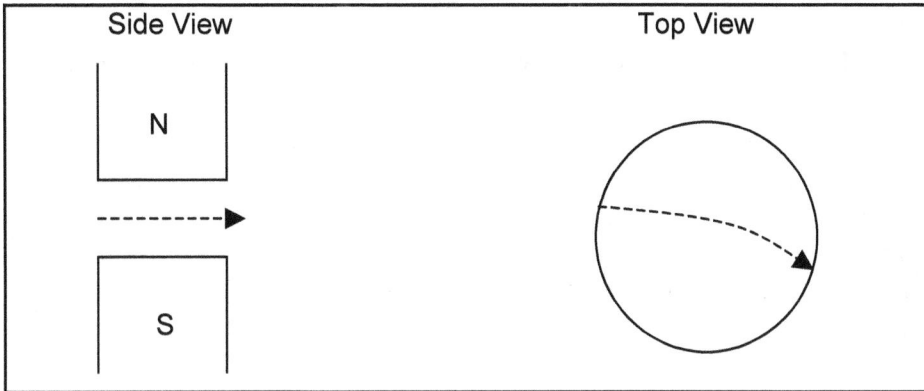

Figure 6.1. Path of a particle between the cylindrical N and S poles of a magnet. The dashed line shows the path of a charged particle: side view showing the particle is not deflected vertically, and top view showing the particle is deflected horizontally to the right.

6.2.3 Spherule Synchrotron Radiation

Accelerating charged particles radiate energy. In the case of a synchrotron of radius ρ the power radiated is

$$P = 2q^2e^2c\ \beta^4/[3\rho^2(1-\beta^2)^2] \qquad (6.6)$$

When $\beta \sim 1$ (relativistic)

$$P \approx [2q^2e^2c/(3\rho^2)]\ (E/mc^2)^4 \qquad (6.7)$$

Since the mass ratio of a u-238 ion to a 0.01 mg spherule is 0.5×10^{-15} the power radiated per atom by an accelerating spherule at a given energy is much less than that of a single U-238 ion:

$$P_s/P_h \approx (q_s/q_h)^2(m_h/m_s)^4 = 1/(92n)^2 \qquad (6.8)$$

or for an element of atomic number A and an n atom spherule

$$P_s/P_h \approx 1/(An)^2 \qquad (6.8a)$$

In the case of a 0.01 mg U-238 spherule the power radiated by the spherule is a factor of 2.7×10^{-35} less than the power radiated by a U-238 ion. Thus the power loss is very much smaller for spherules.

6.2.4 Spherule Synchrotrons are Possible Today

Although turning magnets must be much more powerful and electric rf generators must also be much more powerful, spherule accelerators are viable with present technology or modest extensions thereof. The much lower rate of synchrotron radiation is a very favorable factor.

7. Superluminal Starship Drive

7.1 Benefits of starship R&D – Fusion Power and Advanced Particle Accelerators

The Research and Development of a starship will be a long (30 – 100 years), expensive effort that will be best done by a consortium of technologically advanced nations including the USA, United Kingdom, China, Japan, India and The European Union. An important motivation for this effort is its impact on the research and development efforts for fusion power as well as particle accelerator development.

Figure 7.1 shows the benefit fallout from a starship development effort as envisioned in this book.

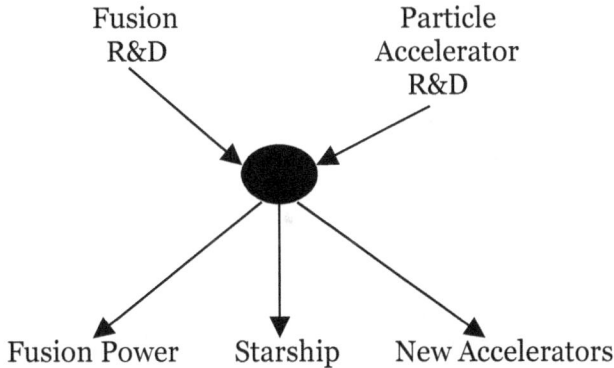

Fusion R&D   Particle Accelerator R&D

Fusion Power   Starship   New Accelerators

Figure 7.1. Benefits from a starship R&D effort accrue in other areas such as fusion power – a vital requirement for the future and new particle accelerators to advance human knowledge.

7.2 Colliding Spherule Starship Thrust

Our starship design will be based on the R&D efforts expended to develop fusion power reactors based on Inertial Confinement Fusion and the extensive experience gained in building ultra-high energy colliding beam accelerators.[44] The steps of the engine cycle are:

1. Highly charged spherules are accelerated in "standard" colliding rings with new, more powerful magnets and rf accelerator modules. Two streams of spherules are diverted by magnets into the collision module as illustrated by Fig. 7.2. They are bent from their circular trajectory to almost the thrust direction.

2.   The about-to-collide spherules are compressed by sets of laser beams similar to those used for inertial confinement fusion to nuclear density in the shape of thin ellipsoids. The long axis of each spheroid is oriented in the "thrust" direction to give a maximal collision surface for the colliding spherules. Each of the colliding spherules are compacted to ultra-high density as if being prepared for fusion (at perhaps twice nuclear density) using an array of powerful lasers or particle beams focussed on each spherule. After compaction the 2 spherules collide in a few fm/c to create a macroscopic fireball. The lasers/particle beams may vaporize part or all of a spherule but the ultra high density of the resulting stream makes that issue irrelevant.

3. The colliding fireballs in the femtometer-size starship "combustion" chamber are confined to the chamber by laser or particle beams on the sides and in front of such strength as to only allow the fireballs to exit to the rear providing a quark-gluon thrust. The strength of the confinement beams must be orders of magnitude stronger than beams used in Inertial Confinement Fusion (ICF) since the density and explosive force is so much more than in ICF devices. The complex velocities of the quarks within the quark-gluon thrust enable complex starship mass and the speed of light to be exceeded.[45]

---

[44] Since the mathematical analysis of the starship propulsion mechanism requires complex computer codes that have yet to be developed we shall only be able to describe starship propulsion in qualitative terms. A mathematical presentation would require a significant programming effort, and extensive experimental input, that is beyond the current state of the art.

[45] The fireball expanding to the rear due to the confining effect of beams assumes an ellipsiodal shape as it exits from the starship. Therefore the real and imaginary parts of the velocities of the expanding fireball are such that the real parts of the quark velocities are perpendicular to the surface of the ellipsoid and the imaginary parts of quark velocities are tangent to the ellipsoid surface. Consequently, the component of the velocity of a quark in the rearward direction will be complex-valued.

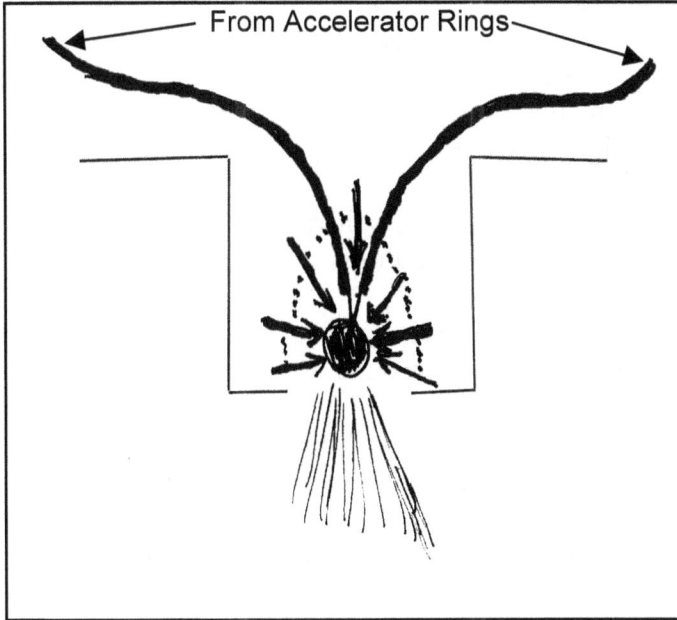

Figure 7.2. Diagram of the collision region of the intersecting spherules. The spherules (thick lines) that are extracted from the rings are thick until they are near the point of intersection where each spherule is compressed by high power lasers or particle beams (not shown) to nuclear density. They then collide generating a fireball that is constrained from expanding by multiple beams except towards the rear from which the fireball of quark-gluon plasma streams to generate the starship thrust. The straight lines enclosing the "combustion chamber" indicate the enclosing sides of the starship hull. The arrows represent some of the laser or particle beams used to enclose the fireball allowing expansion only towards the rear. The "dots" between the arrows indicate that there is likely to be an array of many beams confining the expanding fireball to expansion to the rear of the starship as shown.

4. The thrust should have a high complex velocity of at least the order of c(1 + i), and an expelled mass of perhaps of the order of .001 mg/sec. The thrust will accelerate the starship to a speed well in excess of the speed of light. Since the exhaust speed is so high, small amounts of matter in the thrust will cause the starship speed to increase to high velocities.

5. The energy of the colliding spherule rings should be as high as possible to produce maximal thrust velocity.

The success of the starship engine requires carefully synchronized events at the fm/c level. The effort expended in developing this starship engine will also result in advances in ICF reactors for fusion power and in elementary particle accelerators for deeper studies of elementary particle physics and cosmology. Thus the R&D payoff will be very substantial.

7.3 Fireball Complex Spatial Momentum "Bubble" in Real Space-Time

The fireball created by the high energy collision of heavy ions or heavy atom spherules is substantially different from the collision region in nucleon-nucleon collisions. Fig. 7.3 shows a visualization of the fireball. Inside the fireball quarks and gluons form a perfect fluid and have complex spatial momenta. Outside the fireball the normal real-valued spatial dimensions prevail.

A fireball is created by the collision of heavy ions,[46] and rapidly (in a fm/c or so) becomes a perfect fluid described by thermodynamics. The fireball rapidly expands as described in section 6.1. After 6 – 7 fm/c or so its energy density decreases and the expanded fireball enters the "freeze out" stage where the quarks and gluons in the fireball combine (including quark-antiquark pairs created from the vacuum) to produce "normal particles" such as protons , neutrons, pions, and so on.

---

[46] Other remnants of the colliding ions are also produced.

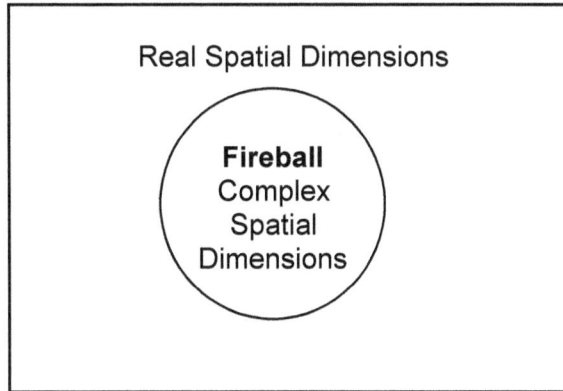

Figure 7.3. A depiction of a quark-gluon fireball in which the quarks and gluons have complex spatial momenta. The fireball is a complex "bubble" within the real space of our experience.

In the case of colliding spherules a much larger fireball is produced if the colliding spherules are first compacted to nuclear density. Each compacted spherule may be viewed as a "super-nucleus."

The surface of a fireball has a certain surface tension – otherwise it would not be approximately spherical in shape (it does have some ellipticity at the 15% - 20% level). However the speed of fireball expansion shows the surface tension is much less than the pressure of expansion and thus may be ignored to leading order.

The existence of surface tension, and the freeze out of ion-ion collision fireballs after 6 – 7 fm/c, does raise the issue of whether the starship thrust experiences a drag due to these effects.

7.4 Freeze Out Stage – Impact on Starship Thrust

The laser or particle beams defining the "combustion chamber" sides confine the expanding fireball to expand only to the rear of the starship and thus provide thrust. The expanding fireball's perfect fluid elongates as shown in Fig. 7.4 and the fluid expands out the rear of the chamber. In a short time period of perhaps a few fm/c freeze out occurs and the fireball "dissipates" with the quarks and gluons combining (together with quark-antiquark pairs excited from the vacuum) to transform into pions, protons, and so on.

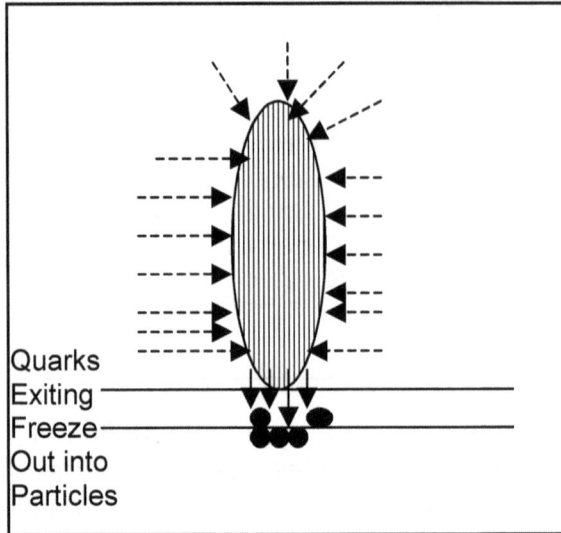

Figure 7.4. Diagram of the expanding, beam guided fireball exiting the "combustion" chamber. Dotted line arrows represent the confining laser or particle beams. The emerging quark-gluon fluid (indicated by the first horizontal line) dissipates (freeze out) and forms "normal" elementary particles (indicated by the second horizontal line).

One may ask if the freeze out, which represents the breakup of the fireball surface, causes a "drag" to occur that reduces the thrust.[47] It appears that the quark thrust, which has a complex velocity, is the true thrust of the engine. The later recombination of the quarks and gluons outside the engine produces normal particles with real velocities. The imaginary part of the quark velocities is subsumed within the produced normal particles. This process of thrust transforming to another form occurs in chemical rockets as well. The molecules in the chemical rocket thrust, which often contains some free radicals, often recombine in the tail of the rocket thrust in a manner analogous to freeze out. The chemical recombination in the thrust tail does not affect rocket performance.

---

[47] Similar drag effects occur in fluid dynamics near surfaces.

7.5 Starship Engine Energy Source

The starship engine that we have designed requires massive amounts of energy. At this point in time the only feasible energy source for starships is nuclear energy. It is reasonable to expect that fusion energy, a more concentrated energy source, will become a reality within the next thirty years. In either case the starship will need an energy source to drive the spherule accelerator rings and associated devices of the engine for periods up to perhaps a few months, then turn off for perhaps many years, and then resume operations for further maneuvers.

In the extreme case of travel to another galaxy, the energy source will need to turn off for up to millions of years of starship time. While the energy source is turned off, a residual "battery" will need to operate to support monitoring the progress of time, activating the startup of the main energy source, and possibly to detect and monitor objects ahead in the line of flight. This battery source may well be a plutonium source similar to those used in current space probes.[48]

The main energy source, if it is a nuclear reactor of some kind, will probably have to be a reactor that is different from current nuclear reactors. The nuclear fuel may have to be suspended in a liquid that becomes concentrated when the reactor is operating and becomes dilute when the reactor is brought to "stop." The dilution factor will determine the rate of nuclear reactions in the reactor. Since the startup process from a battery driven state needs to be gradual due to a "small" battery, it appears the nuclear reactor would be composed of perhaps five reactors of increasing size. The battery starts the smallest reactor by concentrating its nuclear fuel. The smallest reactor then generates the energy to concentrate that fuel, and start the second smallest reactor, and so on until the main reactor is started. At this point the accelerators power up and the starship thrust begins to occur.

If the source of the energy is fusion energy then the startup process might begin in small stages in the boot up of the fusion reaction through fusing larger and larger amounts of (perhaps) ^{3}He with increasingly powerful laser beams a la the tokamak approach.

During the coasting period of a starship, nuclear reactors should be powered down to conserve nuclear fuel. When powering down a

---

[48] We note that a natural nuclear reactor existed in Central Africa for millions of years. (Parenthetical note: Could this be the stimulus for the rapid evolution of species in Africa including early Mankind?)

nuclear reactor based power source the nuclear material (U^{235} or plutonium) (the reactor fuel) residing in the liquid medium of the largest reactor would be diluted to sharply reduce fission reactions to "near zero" using the energy of the next largest reactor. Then this reactor would be similarly powered down by dilution using energy from the third largest reactor and so on until all the nuclear reactors are powered down. The smallest reactor would be powered down by a battery. This battery would retain enough energy to bring the smallest reactor back up after the coasting period ends. Then the reactors would boot up in turn to provide energy to the vehicle. (The battery would be at extremely low temperature during a coasting phase and thus not lose a significant amount of electrical power.)

In the case of a fusion power source a battery could be used to initiate the fusion power. A gradual turnoff process could execute at the start of a coasting phase to bring the fusion process to zero in such a way that the battery could initiate the boot up process for the fusion power source at the end of a coasting period.

7.6 Fuel Consumption on Starships

The high speeds attainable through the quark drive mechanism naturally raise the question of the amount of fuel required to reach those speeds. A major advantage is the high speed of the quarks ejected to produce the starship thrust.

We shall use the equations describing superluminal starship dynamics starting from eq. 3.8 and consider them for the numerical example in section 3.4. In this example, the thrust was generated by ejecting a constant amount of mass per second at high velocity:

$$dm/dt = 1 \text{ gm/sec} \tag{7.1}$$

We can calculate the total mass ejected to obtain a specific starship speed from

$$\Delta m = \int_{t_0}^{t'} dt' \, dm/dt'$$

$$= (t' - t'_0) \times 1 \text{ gm/sec} \tag{7.2}$$

We can relate starship time intervals to earth time intervals so that eq 7.2 becomes

$$\Delta m = (\beta^2 - 1)^{1/2} (t - t_0) \times 1 \text{ gm/sec} \tag{7.3}$$

If $t - t_0$ is approximately one month and $\beta = 5,000$ then

$$\Delta m = 6480 \text{ metric tons} \tag{7.4}$$

for a 100,000 metric ton starship. If $t - t_0$ is approximately one month and $\beta = 30,000$ then

$$\Delta m = 32,400 \text{ metric tons} \tag{7.5}$$

for a 100,000 metric ton starship. Thus the ejected mass in both cases is a respectfully small portion of the total starship mass – especially in comparison to present day chemical rockets.

We conclude that matter consumption for thrust is favorable for quark drive starships.

7.7 Alternate Starship Engine Designs

In an earlier work Blaha (2009b) we proposed an alternate design that had a quark-gluon fluid acceleration ring. It appears that this design is not feasible with current or near term (one hundred years) technology. The reasons are:

1. The quark-gluon fluid ring is not feasible because fireballs cannot be injected and accelerated to form a ring in a few fm/c, which is all the time that is available.

2. A quark-gluon fluid ring would be similar to fusion tokamaks but much more challenging in its requirements for confinement and stability due to the much higher density and temperature of a quark-gluon fluid ring.

For these reasons, the relative simplicity of this new design, and the possibility of using the LHC to build a prototype in the near future, we believe the design proposed in this book is preferable.

8. Building a Prototype Starship Engine on Earth Using the LHC

The LHC (Large Hadron Collider) at CERN in Geneva, Switzerland collides hadrons at center of mass energies up to 7 TeV at present. (See Fig. 8.1 for a diagram of the LHC.) Early results from the ALICE and CMS hadron-hadron collision experiments are available and summarized for our purposes in section 6.1. After the planned series of elementary particle experiments are completed in 6 – 8 years it is reasonable to consider modifying the LHC for spherule-spherule experiments. The 27 km LHC tunnel in which the LHC is housed and other facilities associated with the LHC cost of the order of 60% or more of the total cost. Retooling the LHC and its other facilities using the existing tunnels would yield major cost savings of the order of billions of dollars in the construction of a quark drive prototype as envisioned in the previous chapter. Thus the LHC investment can be leveraged into a starship engine prototype with the additional benefit of stimulating ICF beam containment R&D for fusion power reactors.

Consequently we propose the following series of steps:

1. Verify experimentally that quarks have complex valued velocities as described in the derivation of The Standard Model in Blaha (2010a). Possible experimental tests are suggested therein.

2. If complex valued quark velocities are proven experimentally, then parts of an LHC modified for spherule-spherule collisions could be built in parallel with the elementary particle experimental program.

3. After the particle physics experimental program is completed in perhaps 6 – 8 years, then the LHC can be modified using the apparatus developed in step 2 to implement the starship engine

prototype and a test program initiated. Practically all electronic parts of the LHC depicted in Fig. 8.1 must be modified (or rebuilt) for spherule-spherule collisions. But the in-ground construction of tunnels need not be redone resulting in major cost savings.

4. Upon successful testing of the LHC prototype, the design of a working starship in orbit around the earth can be developed and implemented.

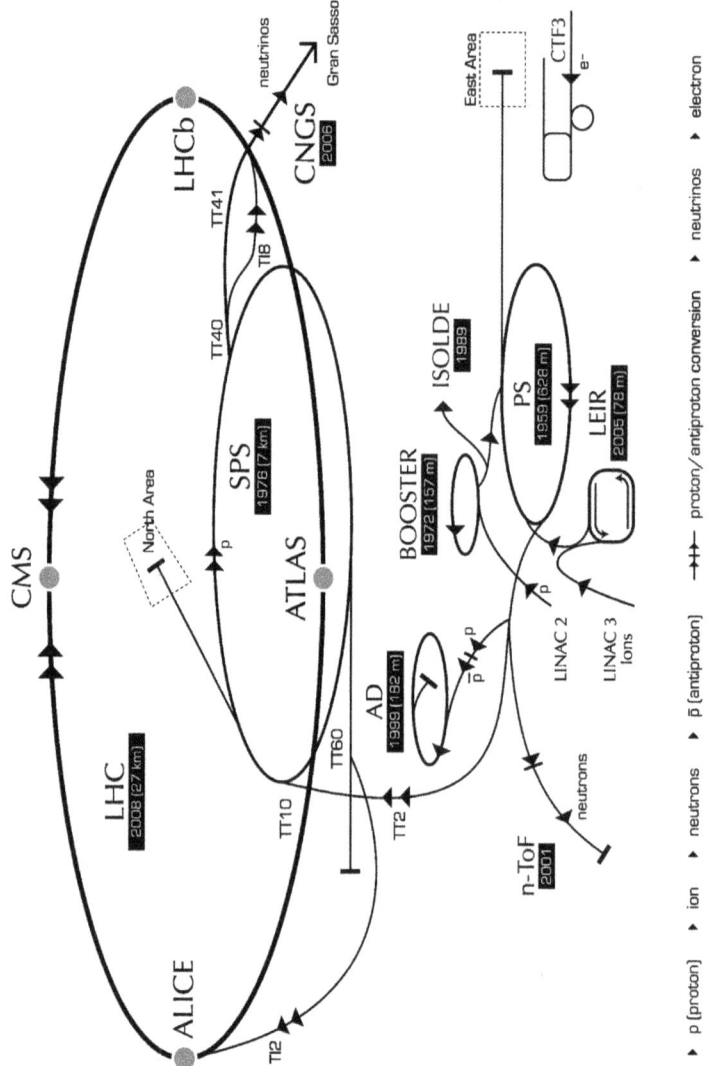

Figure 8.1. The LHC is the large ring in the depicted chain of particle accelerators. The smaller machines progressively help boost particles to the beginning energy required by the large ring. Diagram artist: Christiane Lefèvre, © CERN 2008, Used with the kind permission of CERN.

9. Nuclear Rocket Drives

A starship requires nuclear or fusion engines for local movement within a star system. A starship engine is not intended for "short" distance travel within a solar system or to land on large planets. (It would contain shuttles for that purpose.)

Nuclear rockets would be used for "short" distance travel within solar systems. It would seem reasonable to have four nuclear (fusion) engines arranged in a cluster to provide redundancy for reasons of safety in case one or more engines failed. The exact design of these engines remains to be determined. Preliminary designs do exist from NASA work in this area in the 1950's and 1960's.

Russia has begun a nuclear rocket development program. Russia hopes to use nuclear rockets to explore the planets within this solar system. Nuclear rockets would vastly shorten travel times to Mars and other planets. The starship program could use Russian nuclear rocket technology to create the required nuclear rockets. It is also possible that the United States would revive its nuclear rocket development program of the 1960s.

The experience gained using Russian or American nuclear rockets within our solar system would be an invaluable aid to the development of the starship nuclear engines.

10. Suspended Animation

Suspended animation of a starship crew is not required for travel to nearby stars at five or ten times the speed of light. However for long distance travel to the stars suspended animation is necessary. If the travel to a star at a distance d takes earth time t, then since

$$t = d/v \qquad (10.1)$$

we find that the elapsed time t' on the starship is

$$t_r' = i\gamma(t - \beta d/c) \qquad (2.2)$$
$$= \gamma^{-1}d/v$$
$$\approx d/c$$

using the transformation eq. 2.2. Thus

$$t_r'/t \approx v/c \qquad (10.2)$$

Time progresses faster on a starship moving at speeds must faster than the speed of light by a factor of v/c. For example if $v = 5{,}000c$ then starship time runs 5000 times faster. A 2 month trip from earth's view would be around 1,000 years from the view of the occupants of the starship.

Therefore a practical method of suspended animation must be found. A 4 month round trip to a star at 5,000c would require the starship occupants to be in suspended animation for approximately5000*(1/3 year) = 1667 years – starship time. With suspended animation the crew would be biologically in about the same state at the end of the starship flight as they were at the beginning despite the starship elapsed time of about 1667 years.

10.1 Suspended Animation Research

Many groups are working on various approaches to suspended animation primarily for medical purposes. The development of suspended animation for people is of importance for long range starship travel but it would also be of great benefit in medicine. *It is clear that successful suspended animation necessarily involves lowering body temperature to significantly below 0° C in order to preclude the normal processes of decay to which all flesh is prone. Therefore lowering body temperatures significantly is an absolutely necessary feature of suspended animation.*

When a water-based organism has its temperature lowered the water within it expands causing cells and organs to burst unless some mechanism prevents or avoids the water expansion. Suspended animation studies attempt to find such a mechanism.

The current approaches to suspended animation (which all include lowering body temperature) are:

1. Replacing part or all of the blood in an organism with an "antifreeze" solution that will prevent cells and body tissue from bursting when the temperature is lowered. Revival takes place by raising the temperature of the organism while returning blood to the organism's circulatory system. This approach has been successfully applied to dogs that have been put into suspended animation for three hours. Unfortunately some of the dogs had nerve and coordination problems after revival.[49]

2. An organism can have a chemical injected or absorb a chemical while breathing that will counteract the tendency of water to expand when body temperature is lowered and/or lower the metabolic rate of the organism.

3. NASA and other groups have studied the possibility of placing humans into hibernation. Since hibernating organisms do age – perhaps more slowly – this approach is not true suspended animation.

---

[49] At the University of Pittsburgh's Safar Center for Resuscitation Research.

For the purposes of far reaching starship travel, suspended animation must lower the aging of the human body by a factor of at least 100,000. Given the goal of eventually having a fleet of starships and significant human migration to the stars with livestock and other animals, the mechanism for suspended animation must be relatively simple and safe. Thus approach 2 seems best – injecting or breathing of a chemical(s) that will prevent cell rupture when body temperatures are lowered below 0° C and reduce the metabolic rate to zero. Revival would take place through gradually raising body temperature while diffusing a chemical(s) in the body that counteracts the suspended animation chemical(s).

Developing a successful suspended animation mechanism, that would keep starship occupants alive for up to millions of years, would make travel to far stars in this galaxy and to other galaxies possible.

11. Long-Lived Starship Equipment and Machinery

As we noted earlier time goes far more rapidly on a starship traveling at many times the speed of light. Just as suspended animation is necessary to prolong the lifetime of people, long lived equipment and machinery also is necessary for these long journeys. Typically electronic equipment is engineered for short lifetimes typically of the order of 10 – 20 years. Machinery also is engineered to have a short lifetime rarely exceeding 50 years. An exception to this practice are intergalactic probes launched by NASA that have plutonium power supplies and are designed to last for perhaps hundreds of years.

In the case of starships we see that several life time design goals make sense: a starship with 30 year lifetime equipment for initial exploration of nearby stars, a starship with 200 year lifetime equipment for a wider exploration region, and a starship with 1,000 year lifetime equipment for exploration of a wide nearby region of the galaxy. Ultimately, starships with 1,000,000 year lifetime equipment will be needed for galaxy wide exploration and travel to nearby galaxies if we can reach ultra fast speeds.

For example, if we travel the approximately four light years to the nearest star at a speed of 4c then it will take approximately one year to reach the star in earth time but the starship occupants will experience approximately four years of travel and thus age four years. The components and equipment on the starship will also age four years. For this short trip at "low" speed the time expended on the starship is not excessive.

But suppose we consider a trip to a far part of the galaxy 20,000 light years away at a speed of 5,000c. Then the earth time expended will be four years but the starship time expended will be $4 \times 5,000 = 20,000$ years of starship time. Trips to other galaxies would require millions of years of starship time although the earth time expended would be small. For example, to visit the Andromeda Galaxy 2,000,000 light years away at a speed of 1,000,000c would require two earth years and 2,000,000 starship years.

At present we know that some materials can last millions of years and retain their form albeit perhaps in a chemically changed form. Fossils from many millions of years ago prove it. The challenge is to develop materials and instrumentation for starships that retain their shape and functionality for exceedingly long times. New super strong materials are now being created in the laboratory. For example, large gem quality diamonds can now be made in the laboratory. A major effort to make the required types of materials will undoubtedly prove successful.

Again, there are major collateral benefits in other technology areas to the development of ultra long lasting equipment.

11.1 Robot Guidance and Robot Exploratory Starships

While the occupants of a starship sleep in suspended animation sensors will be required if matter exists in complex space and also to monitor the starship's motion. Computer programs will be needed to guide the starship's motion turning the starship's engines on and off, and regulating the thrust. Again we see that current computer technology would probably suffice for short trips to the stars. But extremely long life stable computers would be required for travel to far stars and galaxies. Creating computers with working lifetimes of up to millions of years is a major research challenge. Since the computers will be in a very low temperature environment the challenge of building long life computers is somewhat reduced.

Prior to manned starships it seems reasonable to send computer-guided, robot starships on round trips to nearby stars to explore as well as test the starship. This period will be similar in spirit to the current probes being sent to Mars and the outer planets.

12. An Exploration Program for the Stars and Galaxies

12.1 How Do We Start?

The Starship Project begins with the development of a working starship engine prototype. The physical structure of the LHC makes it the most economical site for the development and testing of the starship engine prototype.

The Starship Project is rightly an international effort that should be spearheaded by the leading nations. It will be expensive but the payback for a successful effort will be an Open Door to the Universe. The costs can be spread over a number of years. We suggested a thirty development time period in Blaha (2009b). The recently announced NASA Starship Project suggests a one hundred year time frame. With these time frames the costs should be roughly of the order of ten billion dollars a year.

Chapters 1 and 8 provide a plan of action to design and build the starship.

12.2 Exploration Phase

There are about 100 stars within a 21 light year radius from earth. The proposed starship with its high superluminal speed capabilities will enable trips to stars within 100 light years of earth with relatively short travel times. Astronomers will soon be in a position to develop a priority list of stars to visit with planets that could provide homes for Mankind. The starship could then perhaps visit and explore the top ten stars on the priority list. If truly earth-like planets are found then the construction of a fleet of starships and the creation of initial colonization outposts can then begin.

12.3 Seeing and Navigating through the Cosmos on Superluminal Starships

The view of the universe that a starship crew sees when the starship is traveling faster than the speed of light is very different from the view of a spaceship traveling at low speeds of a few tens of miles per second.

As Mallove (1989) points out[50] an observer on a starship traveling at a relativistic speed near but below the speed of light will see the visible stars and galaxies compressed to within a cone in the direction of the starship (Fig. 12.1). The cone gets narrower as the speed of light is approached due to aberration and in the limit as the speed approaches the speed of light becomes a point directly ahead of the starship.

The relativistic equation for aberration is

$$\cos \theta' = (\cos \theta + \beta)/(1 + \beta \cos \theta) \qquad (12.1)$$

where θ is the angle of a star (or galaxy) relative to the starship's direction of motion as measured in the earth coordinate system, and θ' is the angle of a star (or galaxy) relative to the starship's direction of motion as measured in the starship's coordinate system.

The inverse relation is

$$\cos \theta = (\cos \theta' - \beta)/(1 - \beta \cos \theta') \qquad (12.2)$$

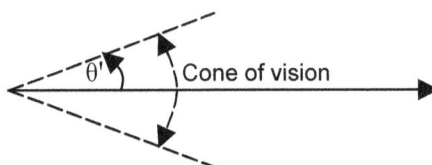

Figure 12.1. Cone of vision around direction of starship motion in the starship coordinate system with the angle θ' determined by eq. 12.1 for sublight starship speeds.

---

[50] pp. 182-185. They reference the scientific papers that are the basis of his description.

12.3.1 Sublight Case: $\beta < 1$

As $\beta \rightarrow 1$ (the speed of light) eq. 12.1 indicates $\theta' \rightarrow 0°$ showing the entire view of the universe is compressed to the exactly forward direction. Fig. 12.1 shows the cone of visibility for a spaceship traveling near the speed of light at perhaps .6c - .9c. The cone angle θ' satisfies

$$\cos \theta' > \beta \qquad (12.3)$$

The rest of the field of view of the starship is total blackness except for the point in the directly rearward direction ($\theta' = 180°$) where an object at $\theta = 180°$ can be seen as a point.

12.3.2 Superluminal Case: $\beta > 1$

For $\beta > 1$ eqs. 12.1 and 12.2 still hold and there is a cone of visibility similar to that depicted in Fig. 12.1 . However the cone angle θ' for superluminal speeds, $\beta > 1$, satisfies the relation

$$\cos \theta' > 1/\beta \qquad (12.4)$$

The rest of the field of view of the starship is total blackness, as in the sub-light speed case, except the point in the directly rearward direction ($\theta' = 180°$). We note that as β gets very large the cone of visibility becomes larger. At $\beta = \infty$ the cone of visibility becomes the angular region between $\theta' = 0°$ and $\theta' = 90°$ (the forward hemisphere).

12.3.4 Superluminal Starship Visibility

As a result of the above considerations, visual navigation at high superluminal speeds becomes difficult although one can conceive of electronic imaging that "undoes" the effects of aberration and enables visual navigation.

A further problem is the location of a destination. If we send a starship from the earth to a star, for example 30 light years away, we have to project the location of the star at the time the starship arrives based on the star's current motion as determined by earth observation. If the motion of the star is modified by the gravitation effects of other nearby stars during the 30 years that the light from the star was traveling to earth, or if the star's motion is not accurately determined, a starship could arrive at a point that is some distance from the star.

Thus navigation to a destination is a significant issue.

12.3.5 Effect of Doppler Shift at Superluminal Speeds

A starship traveling at relativistic sublight speeds will see stars having their color changed significantly due to the Doppler Shift effect. At superluminal speeds the Doppler Shift will also have a significant effect that will change the view of starship occupants. This issue is again surmountable if we use electronic imaging techniques to "undo" the Doppler shift and thus display stars as they normally look in the visible human frequency range.

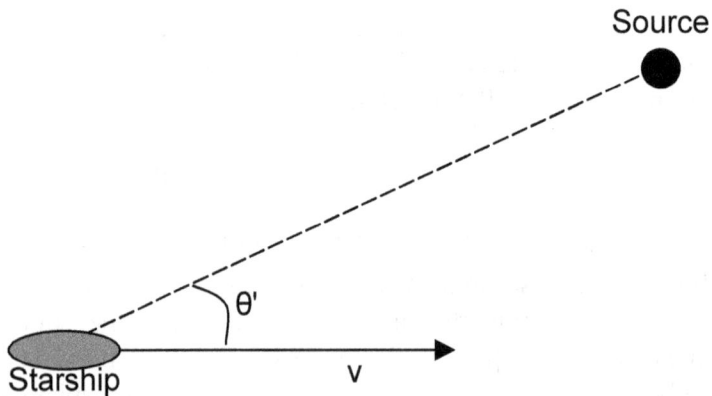

Figure 12.2. The angle of the source θ' with respect to the starship's velocity **v**.

The relativistic Doppler shift for sublight speeds of a light wave of frequency v is given by

$$v = v_0(1 - \beta^2)^{\frac{1}{2}}/(1 - \beta \cos \theta') \qquad (12.5)$$

where v_0 is the frequency of the light emitted by the source and θ' is the angle of the source relative to the starship's velocity (Fig. 12.2). For superluminal speeds the Doppler shift is

$$v = v_0(\beta^2 - 1)^{\frac{1}{2}}/(\beta \cos \theta' - 1) \qquad (12.6)$$

This can be seen by considering an electromagnetic plan wave, which is a combination of

$$\cos[(k{\cdot}x - vt)/2\pi] \quad \text{and} \quad \sin[(k{\cdot}x - vt)/2\pi] \qquad (12.7)$$

Upon transforming from the earth coordinate system, for example, to a coordinate system moving in the x-direction at a speed faster than light both the energy (v' up to a constant) and the time t' acquire a factor of i (that cancel each other) so eq. 12.6 is the correct frequency in the superluminal frame. The sign of the frequency is always positive by convention due to the form of electromagnetic waves and eq. 12.4 dictates the form of the denominator in eq. 12.6.

For large $\beta \gg 1$ eq. 12.6 becomes approximately

$$v \approx v_0/\cos \theta' \qquad (12.8)$$

In the forward direction $\theta' = 0$ the Doppler shift goes to zero. Due to eq. 12.4 the maximum value of the Doppler shift for large β in the field of vision is

$$v \approx \beta v_0 \qquad (12.9)$$

So the "wide" angle electromagnetic waves are shifted to large frequency.

Eq. 12.6, and the discussion that follows, suggests that frequency shifts will be substantial for extremely fast starships. The result will be a distorted view of the universe. However electronic imaging techniques can be implemented to restore the view that humans would normally see. Thus the combined effects of aberration and the Doppler shift on the view of space from a starship bridge can be electronically corrected to give a "normal" view of the oncoming space.

12.4 Possible Encounters with Aliens

The possibility that a starship will encounter other forms of life when they visit planets in other solar systems is very real. There are three categories of other forms of life. Each category requires that a different procedure be followed by the starship crew. The categories and the necessary reactions of the starship crew are:

1. Non-intelligent non-carbon based life – Life of this sort, such as silicon-based life, does not appear to pose a health hazard to humans. However, it is possible that animal-like predators would be a threat.

2. Non-intelligent carbon based life - Life of this sort could pose a health hazard to humans. It is also possible that animal-like predators would be a threat.

3. Intelligent life – Both a threat and an opportunity. Dealing with intelligent life is a complex issue. We have addressed this issue as a meeting of civilizations in Blaha (2010b) and earlier books on civilizations. The following subsections address specific issues of first meeting intelligent aliens.

12.4.1 The Problem of Communication

Communication with other forms of intelligent life will be a major issue. The problems that we have encountered on earth in communicating with other intelligent and semi-intelligent species such as elephants, whales, dolphins, monkeys and so on proves this assertion. After decades of study our ability to communicate with any other species on earth is rudimentary.

If we should encounter aliens on some other planet our attempts to communicate will most likely first begin with the "naming of things" – concrete visible objects. The transition to abstract concepts such as emotions (love, hate, friendship, and so on) could be difficult since it is possible they have concepts in their language that are unknown to us, and some of our concepts may well be unknown to them. An example of differing concept levels is the concept of love – one word in the English language but hundreds of words with different nuances in Greek.

The expectation that mathematics or science can offer a common ground of discourse may well be met but with some major difficulty as mathematical and scientific concepts are often subtle and distinctions between concepts can prove difficult to discern. A simple example is quantum mechanics, which after almost 100 years, is still subject to disputes over concepts and meanings.

Thus communication with aliens – if possible – will be difficult at best.

12.4.2 Other Issues in Alien Encounters

Other issues, which some commentators feel are significant, are the possibility of alien hostility and the possibility that, in finding them, they might find earth and perhaps attack. Clearly, a starship should be armed with an appropriate array of weaponry including nuclear weapons to deal hostile encounters. Whether a starship's travel to a star can be traced back to its origin on earth is an open question that depends on the alien's level of technology. Switching to nuclear engines as one approaches a star system may serve to deflect this issue.

In any case one can hope that any intelligent life we meet amongst the stars will be friendly. It appears, on the basis of chance and "luck", that there is at least an *a priori* 50-50 probability that they will be friendly.

Appendix A. The Earth and the Need for Solar System Exploration

A1 The State of 21st Century Earth

A1.1 21st Century Earth

The year 2011 is a remarkable year – not because of the armed conflicts that are taking place in various parts of the world – but because we see for the first time that the earth is a "zero sum game."[51] The gains experienced by one country were counterbalanced by the losses of other countries. In the past decade the United States has "lost" a substantial amount of its capital through foreign trade to Asian and European countries. The prosperity of these countries has increased. Now the United States is facing massive economic problems because of the loss of many industries and the outflow of its wealth. A consumer country without a strong manufacturing base becomes an economic bleeding corpse.

The growth of prosperity in Asia and Europe is a welcome event but it has become clear that much of it was at the expense of the United States and other wealthy countries. Most people hope that all the countries in the world will eventually reach a common high standard of living not unlike that which the United States has enjoyed in recent decades on average. Such a world would have much less reason for warfare and a world at peace would be a likely prospect.

However this dream cannot be accomplished for several reasons. Given the world's population of six billion and the fact that the United States consumes 40% of the world's production while having only about

---

[51] A "zero sum game" is a game in which one person's gains are counterbalanced with another person's losses. Thus the sum of all players "chips" remains constant and the change in the total sum as the game progresses is zero. The commodities markets (in gold, grains, pork bellies, currencies and so on) are considered to be zero sum games. The gains (profits) of one trader are the losses of another trader(s). Thus the total sum of all traders' investments in a commodities market is unchanging.

5% of the world's population, world production would have to increase by a factor of roughly 8 for the average world standard of living to be that of the US. Eight times more food, eight times more housing, eight times more industry, and so on. Clearly, an impossible task for the earth to achieve and sustain.

Now we must confront the fact that the world today is at its environmental limits. The degradation of the environment at today's production levels is obvious: water and land are increasing polluted (poisoned). In particular, the oceans are being polluted with heavy metals, plastics, sludge and so on. We are entering a period of man-made global warming, increasing air pollution, increasing birth defects and educational disabilities, and other signs of an overstressed and precipitously declining environment. As the population grows to nine billion in 2050 we can expect to see conditions worsen.

And the terrible fact is that we cannot change these trends without massive efforts that will cause a major lowering of the standard of living for all nations. *These facts "everybody" feels but political leaders in every country cannot voice them without causing major population unrest.* Can the leaders of China or India say they can never reach the American standard of living without causing their peoples to lose hope and perhaps turn to unscrupulous demagogues who promise anything to achieve power? Can an American president say that universal lifetime health care, while maintaining current high standards of medical quality, is not economically feasible over the long term?

Clearly world leaders avoid stating the realities of the world situation and put in place "programs of hope" that promise eventual benefits but in reality cannot lead to the nirvanas they boldly promise.

The most significant and most understated problem facing the earth is the need to balance production with effective waste disposal. Manufacturing in all its facets is producing products, which often require waste disposal methods whose costs are comparable to the costs of their production.

A1.2 Zero Sum World Economy

Until recently growth and development were the engines of prosperity. Now we must change to a zero sum economy where "*every dollar spent on production must be matched by a dollar spent on recycling and environmental protection.*"

Further the Chinese program of population limitation should be taken up by the other nations of the world in an appropriate form. The western nations of Europe and North America have a stable or declining population if the immigrant population is not counted. The countries in Asia, South America, Central America, and Africa need to strongly encourage population limitation with the spirit of the Chinese model but not necessarily with the methods of the Chinese government.

An eventual downsizing of the earth's population to about one billion appears to be required if we wish a high standard of living for all combined with a clean, stable world environment.

A1.3 The Effect of a Declining Population

Downsizing the population of a nation or a planet is an undertaking fraught with dangers. Japan is experiencing some of those difficulties as its population declines, and ages. Economic problems include the cost of supporting an aging population. And the major burden falls on the working, youthful part of the population whose percentage of the total population is decreasing.

There is a more subtle cultural issue associated with aging. The creativity and cultural progress generated by the more youthful part of the population is less strong as the relative proportion of younger people declines.

In the case of our world, the population will grow from the current six billion to nine billion in 2050 (a figure which cannot be changed except by drastic events). Then the population must decrease to about one billion perhaps over one hundred and fifty to two hundred years to achieve an eventual, maintainable, high standard of living for all. One can only expect that near totalitarian means will have to be used to achieve this goal. The result of this plan will be to profoundly affect the world's economy and the world's ability to engage in major projects such as space exploration and scientific research.

At present large scale scientific and space projects are carried on by nations with large populations with the economic wherewithal to pay for these projects. Large-scale projects cost little for large countries on a per capita basis. A small population necessarily will have a "small" overall economy and will not be able to finance major projects such as space exploration at a low per capita cost in the manner that we see currently unfolding in the major US, EU, Chinese, and Indian space programs.

Lastly, if the earth's population is reduced to about one billion it is possible that "overshoot" may occur and the population might further decline to unacceptable levels. A historical example of this possibility is the latter phases of the Roman Empire when many areas including Italy experienced large declines in population. Augustus Caesar noting the decline in birth rates amongst the Roman population started the practice of paying bonuses to parents for having children. More recently, France, Germany, and other modern European countries have experienced continuing population declines.

Thus it seems best to make major space efforts now when the world can most afford it rather than make minor showcase efforts—an accurate description of current efforts.

A1.4 An Alternative: Beehive Earth

If we fail to begin major efforts soon then one can expect the earth to eventually have a petrified civilization that will ultimately lead to a "Beehive-Earth" with a stratified society with little or no hope for the future.

With the hope for a better future destroyed it is possible that the world faces a period of fragmentation, and perhaps warfare, competing for resources and clean water.

A1.5 Environmental Protection will Never Completely Work

There is a relatively strong conservation movement in the United States and in other countries around the world. While this movement will achieve victories in the struggle to protect the environment from further deterioration, the shear size of earth's population makes it impossible to prevent environmental deterioration now and for the foreseeable future.

Much of the world is in a growth phase currently, particularly India, China and Southeast Asia. The result of this growth phase is increasing pollution. Eventually an equilibrium point will be reached where growth will give way to environmental maintenance and recovery. The recycling of waste materials will in itself become a major task for the world's population. The consequence can only be a decline in the world's standard of living.

A1.6 Hope: Colonize Space

Approximately twelve thousand years ago the major ice ages ended and the scattered parts of Mankind began to develop societies in various parts of the world. After domesticating animals and plants the population density of Mankind, and its wealth, reached a point where civilizations became possible about 3000 - 4000 BC. Interestingly civilizations[52] appeared more or less simultaneously on the continents of Asia, Africa and South America.[53]

THE EXPANSION OF CIVILIZATIONS

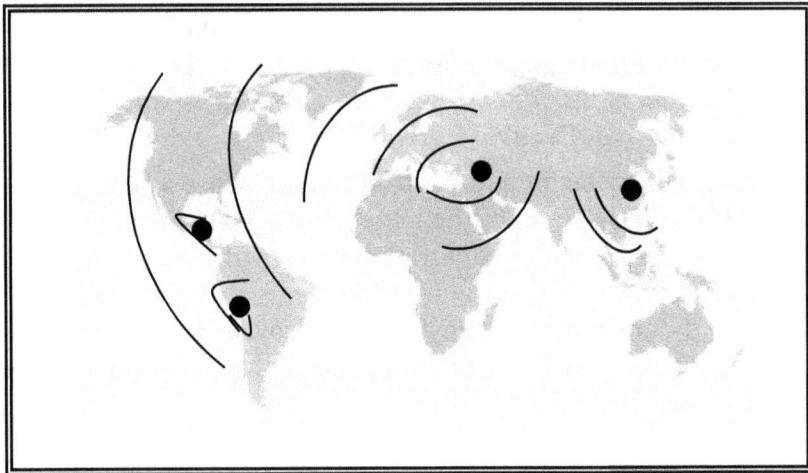

Figure A1.1. Some mushroom rings of civilizations.

Since the first civilizations, Mankind has expanded enormously in population, and in culture and technology.[54] We now have a world

---

[52] Much of the material in chapter 0 and chapter 1 first appeared in Blaha (2006).

[53] In Asia the Yellow River civilization of China, the Mideast civilizations, Egyptian civilization, and the Caral civilization of Peru (recently discovered by Dr. Ruth Shady) all appearing within several hundred years of 3,000 BC. Both the Caral civilization of Peru and Egyptian civilization built great pyramids on a similar scale.

[54] It is this author's opinion that the somewhat astounding advances in Science and Technology—particularly in the past one hundred years—are actually only a beginning and that continued progress in these areas will lead future generations to look upon the present state of these fields of

teeming with people and an explosive growth in communications (satellite communications, the Internet and so on) that seems to be suggest that a world civilization is in the making. Together with that growth we clearly see that Mankind is pushing the limits of the earth's ability to sustain it. The land, air and oceans are undeniably polluted. Since we have reached the limits of the earth the only viable option is to expand into space to new virgin planets.

The first question that we must address is "Why?" We do not presently appear to have any great practical reason to reach out from the earth.[55] And the growth of the human population, from 6 billion or so individuals to many more billions through the colonization of nearby moons and planets in space, is not, in itself, appealing. The answer seems to lie in an old truism, "What does not grow, decays." Thus there appears to be an inherent need for new ground for the continued progress of Mankind. That is the best answer to the "Why?" of space colonization. Until the 20th century, there was always new ground for the growth of civilizations. As we have seen earlier, civilizations radiated outwards from their original locations: in South America, Anatolia, the Mid-East, India, China, Africa and so on. Those civilizations, and parts of civilizations, which were at the leading edge of growth – culturally and technologically – were on the "new ground" for the most part. We call this the *mushroom ring* effect. Civilizations grow outward in an expanding cultural and geographic ring. We have seen these rings of growth in the evolution of Chinese civilization from Sinic civilization, and the expansion of Hellenic civilization from Greece to Italy, the far shores of the Mediterranean, Western Europe, and eventually the Americas; and also into southwestern Asia and India. Other civilizations such as Islamic civilization also show significant mushroom ring effects.

There are easy explanations of the growth differential between "old ground" and "new ground." New ground often tends to be more

---

endeavor as primitive and barely a start. The exploration of the possibilities of the human mind is also in a very preliminary stage in the author's view.

[55] The other planets, moons and planetoids do not contain any exotic elements or chemicals that are not found on earth or synthesizable on earth. Excepting possibly for Astronomy, the search for extraterrestrial life, and certain solid state physics experiments that require zero gravity, earth-based experiments are as feasible as off-earth experiments and usually substantially less expensive.

fertile, has not had the environmental destruction of old ground,[56] and tends to attract the more daring, innovative segments of the population who interact to produce new solutions for the issues facing humanity.

Today, in the absence of new ground, there appear to be three ways in which Mankind can grow: the use of technology to restore the environment and lower the needs of Mankind for raw materials making the earth more habitable;[57] an expansion into space – a societal challenge that could bring about a major advance in civilization; and an expansion of our use and knowledge of the human mind to achieve a higher level of culture. The third possibility does not seem to be practical for the mass of humanity and would lead to a civilization of physical self-denial amongst growing environmental disaster. Therefore we will not consider this possibility further.

A combination of the first two possibilities seems to provide the optimal answer: improve the earth and go into space.

A1.7 Space and Technology as "The New Land"

One possible strategy for achieving a good standard of living for the bulk of the world's population is miniaturization. Technology has made great strides in electronics and computers to make smaller and smaller components and gadgets.

Unfortunately there are certain items: homes, cars, and necessarily human-sized items that cannot be miniaturized. As a result technology can only partially reduce the resource needs of the world's population.

Thus earth-based technology cannot solve the world's problems.

A massive expansion into space is the only "humane" solution for the earth. The requirements for this expansion are the construction of a large fleet of spaceships that can transport large numbers of people –

---

[56] Among the few places where human occupation has apparently improved the environment are certain South Pacific islands were the inhabitants have developed farming practices that have improved the fertility of the soil. Israel has restored the fertility of much of its farmland through innovative farming techniques. Egypt also has restored parts of its northwest (south of Alexandria) using water from the Aswan Dam and innovative farming techniques.

[57] The growing miniaturization in electronics and computer equipment, and possibly nanotechnology, could make it possible for humanity to live well with lesser use of resources. Freeing the massive dependence of industry on water can make more water available for agriculture in desert areas. Genetic engineering could lead to crops that feed more people and require less water and fertilizer. New technology might be able to reverse much of the environmental degradation that has even polluted the (until recently) "boundless" oceans.

particularly to Mars. The technology of suspended animation must be developed so colonists can sleep on their way to Mars. And, firstly, and most importantly, Mars must be upgraded to a human livable planet through bombardment with water asteroids from the asteroid belt and oxygen generation using genetically engineered plant life. Recent studies have shown Mars already possesses large quantities of water, carbon dioxide and iron (for construction). Nuclear reactors to provide power for mining and construction can be widely used on Mars in the absence of any significant population. So the earth has a solution available if it grasps the opportunity.

This solution cannot solve the earth's overpopulation and pollution problems but it can ensure the spread of humanity to nearby moons and planets so that humanity can expand and thrive in new environments. These new environments will stimulate the expansion of human endeavors to higher levels of achievement. Earth is the birthplace and nursery. The universe is the college of Mankind.

A1.8 Why the Urgent Need for Space Colonization Now?

Today the nations of the earth have the resources necessary to launch a major space effort. At the time of this writing (Winter, 2008) the slowdown in the world's economy is causing major unemployment and a decline in consumer demand. Major nations such as China and the United States have announced large public works programs to build and rebuild their infrastructure: roads, bridges, dams, and so on.

They should also consider building spaceports, which are few in number, and factories and support facilities for space rockets. These future-looking projects would furnish employment, and create an infrastructure for space travel. Then a series of manned flight efforts to reach Mars and the asteroid belt, and begin the transformation of Mars should be started. An effort of this magnitude in itself would provide a major boost to the world economy and perhaps redirect us from weapons and armaments to constructing a greater future for Mankind.

Should this opportunity be missed, or begun in a slow fashion, it is possible that the needed resources may not be available fifty or one hundred years from now.

The earth is becoming much poorer.

A2 Stages of Solar System Space Exploration

Numerous proposals have been prepared for the exploration and colonization of nearby moons and planets as well as for large space stations.[58] While many of these are sensible and feasible it appears that the full utilization of available technology and resources has not been properly considered in many of them. Partly, this is due to "emotional" commitments to certain technologies. Partly this is due to the comfort of using previous successful approaches.

Our proposed plan is based on multiple technologies that are suited for each stage of space flight. Because of the success of the American program to place a man on the moon directly from the earth we tend to ignore the fact that it is generally more sensible to engage in space flight in stages with each stage most efficiently adapted to the needs and conditions of that stage.

The solar system stages as we see them are:

1. Earth to Earth orbit
   Mechanism: Chemical Rockets, Space Guns, and Rail Guns
2. Earth orbit to moon orbit
   Mechanism: Chemical Rockets
3. Moon orbit to/from a moon base
   Mechanism: Nuclear Rockets, and Rail Guns
4. Moon orbit to Mars or moon orbit to the asteroid belt
   Mechanism: Nuclear Rockets
5. Eventually, Moon orbit to Venus, Mercury, the asteroid belt, and the moons of Jupiter and Saturn
   Mechanism: Nuclear Rockets

The latter part of this book is devoted to travel to the stars. A new technology is needed for star travel in a meaningful way. Most of the proposed methods are at best "token" travel without significant benefits for Mankind. These methods typically require enormous resources beyond the scope of earth's available resources for the foreseeable future.[59] We will propose a new method for star travel based on an

---

[58] Various approaches are discussed by Zubrin (2000) and Freeman (2009). The author feels the approach proposed here is the most feasible.
[59] See Matloff (1989) for a detailed discussion of the requirements of various proposals.

extension of Special Relativity to faster-than-light travel. Our approach leads to potentially enormous starship speeds. We considered examples of speeds of 5,000 and 30,000 times the speed of light. Much higher speeds are also possible. These speeds reduce star travel and even galaxy travel times to days and months – not generations. The Andromeda galaxy is "just around the corner" in our starships.

A3 From Earth to Earth Orbit

The relatively strong gravity of the earth has locked Mankind to the planet until the 1950's. The only viable method to escape the earth that has been used since the 1950's was the use of ever-larger, chemical propellant rockets. Rockets have enabled us to reach as far as the moon with manned space flights and as far as the edge of the solar system with unmanned probes.

From the 1960's through the 1980's other possible solutions were considered including nuclear rockets and "space guns." These approaches were not implemented for a variety of reasons discussed below.

Recently other approaches, and variations on rocket technology, have been considered by private companies as well as by the US, Russia, China, Japan and India.

In this appendix we will outline some of the approaches to escaping the earth's surface to space. The reader is directed to Matloff (1989) and Freeman (2009) for more detailed discussions.

A3.1 Chemical Rockets

Chemical rockets and modified "rocket planes" are prohibitively expensive if one wishes to move people and equipment in bulk to space and/or earth orbit. Rockets restrict travel to a few essential personnel, and equipment for satellite communications and scientific experiments. The initial steps in rockets to space were based on solid fuel rockets that use ammonium perchlorate as a fuel component. Ammonium perchlorate rockets are not only extremely expensive but also are environmentally dangerous.[60] Large liquid fuel rockets are now being developed that use

---

[60] Perchlorates are neurotoxins that have been implicated in health problems such as tumor growth on the thyroid gland, mental retardation birth defects, and learning disabilities in children. Perchlorates have been found in trace amounts worldwide. Areas near most of the 12,000 military installations and rocket fuel plants in the US exceed the EPA standard of 24.5 parts per billion in

hydrogen and oxygen as propellants. Hydrogen and oxygen combine to produce water – environmentally safe. But the production of hydrogen requires large amounts of electricity, which currently is generated by coal-fired plants. So hydrogen fuel in vast quantities has a significant environmental impact. Despite being environmentally safe liquid fuel rockets require so much energy to produce that they cannot provide "mass transit" to space.

A3.2 Nuclear Rockets

Nuclear rockets are significantly more economical than chemical rockets. However, the possibility of an accident on a nuclear rocket has led space agencies to conclude the risks of nuclear rockets are too high for use for transit from the earth's surface to space. The Chernobyl disaster shows the impact of a nuclear rocket accident in flight could be a calamity for a large area of the earth's surface.

However, nuclear rockets can play an important role in travel in the solar system so we will consider in section A5.

A3.3 Chemical Space "Guns"

Another approach to large-scale travel into space is through the use of "space guns" of the type proposed by Jules Verne. It is a little known fact that the German big guns of World War I (the approximately 100 foot long Big Bertha and a larger gun) that bombarded Paris from a distance of 80+ miles sent their 100+ lb shells as high as 80 – 90 miles above the earth to the very edge of space. The shells had a speed of 1 mile per second as they emerged from the gun's barrel. The German big guns were exceeded by the guns developed in the HARP program led by Jerry Bull 1960's.[61]

Space guns are interesting in the light of the fact that altitudes of about 100 miles are viewed as "Near Space" and that today many space satellites circle the earth at 200+ mile altitudes in "low earth orbit". Space guns can put objects into Near Space. The price to shoot one

---

drinking water by a **factor** of 30,000. Perchlorates appear in 93% of US lettuce and milk, and 97% of US breast milk. Senator Feinstein (CA) has called this situation the "US Rocket Fuel Pollution Scandal" and pressed for strong congressional action.

[61] I am grateful to Dr. Mitat A. Birkan, Program Manager, Space Propulsion and Power, AFOSR/NA for providing this information as well as other details discussed later on chemical space guns.

kilogram (2.2 pounds) at a muzzle speed of 1.6 km/sec is about one kilogram of the best chemical propellant. In comparison a rocket would use many times more propellant to send a kilogram up eighty miles.

A single or multi-stage rocket, with or without boosters, uses over 99% of its weight as fuel to send a payload into space. The fuel is, for the most part, used to propel itself (the fuel) off the ground at an ever-increasing speed into space. In contrast, the propellant for a gun propels the projectile and a fraction of the propellant between the point of burn of the powder charge and the projectile. Thus a sufficiently large gun could efficiently put a payload up eighty miles into Near Space because the propellant propels the payload, and not fuel or a rocket casing. If the space gun is scaled up to send 500 kg payloads into space then we have an effective mechanism to send large amounts of material into space in 500 kg chunks using about 500 kg of propellant per payload.[62]

When the payload reaches eighty km or so then there are two possibilities. The payload might contain a small rocket to put it into a higher earth orbit or to send it to a space station. Or there could be a "scooper" vehicle circling the earth that could scoop up the payload and deliver it to a space station at a higher altitude.[63]

The fabrication of space guns is well within our technology. The major drawback to space guns is the deterioration of the wall of the gun barrel with repeated use. However, the walls can be resurfaced – presumably at a much lower cost than replacing a throwaway rocket or refurbishing a used space shuttle. The other drawback is the design and construction of scooper vehicles. This does not appear to be a significant problem since constantly circling vehicles using atmospheric bounce were designed in the 1950's by NASA although never built.

A3.4 Rail Guns

A *rail gun* is an electrical device that accelerates a conducting projectile along a pair of metal rails. Railguns have two sliding or rolling

---

[62] Dr. Mitat points out that the amount of propellant increases rapidly with muzzle velocity and that at over 2 km/s the required propellant mass is more than three times as large as the payload.

[63] The concept of a vehicle constantly circling the earth at a range of altitudes and "bouncing" off the earth's atmosphere to reach higher altitudes was developed by German scientists in World War II and studied by American scientists after the war. This type of vehicle could be used as a scooper vehicle for payloads shot into Near Space.

contacts that enable a large electric current to pass through the projectile. The rails generate a very strong magnetic fields that causes the projectile conducting the current to be rapidly accelerated to a high speed.

Chemical propellant guns have been created that have reached speeds of 2.6 km/sec. Rail guns have accelerated projectiles to speeds over 6 km/sec in laboratory experiments. Thus a rail gun track running up the side of a high mountain can accelerate a payload to a speed nearly half the escape velocity of the earth (11.2 km/sec).[64] In addition a rail gun does not have the problem of hot gases, corrosion, and barrel deterioration of chemical space guns although rail erosion is a problem. Replacing rails is generally easier than repairing gun barrels.

They also do have the problem of requiring sizeable amounts of electricity. Various laboratories around the world are engaged in rail gun research.[65]

If rail guns prove feasible and economical they could provide a means to send 100 kg (or greater) payloads into space for pick up by a scooper vehicle.

A3.5 Conclusion

Rockets, space guns, and rail guns offer mechanisms to send materials in bulk into space as payloads that can be assembled into a large space station or set of space stations.

These space stations can, in turn, be the terminals for travel back and forth to the moon or planets.

A4. Earth Orbit to Moon Orbit and Moon Base

A4.1 Moon Base Development Phase

After the establishment of a sizeable, perhaps self-sustaining, space station with major stores of rocket fuel, an organized colonization of the moon becomes possible using a combination of conventional rockets and nuclear rockets.

One possibility would be for a chemical rocket to be assembled at the space station and linked to a nuclear rocket also assembled at the

---

64 In contrast the escape velocity from the surface of the moon is 2.4 km/sec.
65 One interesting study of the use of rail guns to assist rocket acceleration is described in Uranga et al "Rocket Performance Analysis Using Electrodynamic Launch Assist", Proceedings of the 43rd AIAA Aerospace Sciences Meeting (January, 2005, Reno, Nevada).

space station. The chemical rocket would carry both itself and the nuclear rocket to the vicinity of the moon – in particular, to just beyond the point where the earth's gravity and the moon's gravity are equal and opposite. Then the nuclear rocket could detach and safely use its nuclear power to establish an orbit around the moon. The nuclear rocket would be beyond the earth's pull and thus the danger of a crash of a nuclear rocket with the reactor operating on earth would be avoided.

The chemical rocket could continue on to land on the moon and be recycled into part of a moon base. The combination of many rockets suitably modified and joined could form a sizeable initial moon installation. At a later stage the moon's crust could be mined, and hydrogen, oxygen, water, iron, titanium, aluminum uranium[66] and other metals could be refined and used for construction purposes, and to make nuclear spaceships that could take advantage of the moon's weaker gravity compared to the earth.

The nuclear rockets orbiting the moon could form a flotilla of ships to establish transportation to Mars and beyond.

A4.2 Moon Colonization

The moon has water, metals and other materials, as well as abundant solar energy to make a moon colony a viable option. The development of mining and manufacturing industries on the moon to produce spacecraft would eventually be more economical than transporting them from earth with an escape velocity almost triple that of the moon.

Since there are many studies and articles on the details of various development proposals we will not discuss a detailed plan here. The reader can find possible plans elsewhere.

A4.3 Mature Nuclear Transportation

The engineering of nuclear rockets can be expected to follow a pattern similar to the improvements in airplanes and jets seen in the twentieth century. Through trial and error, and the analysis of mishaps, nuclear rockets will be improved to the point where they will eventually

---

[66] The Japanese Kaguya spacecraft named SELENE for "Selenological and Engineering Explorer" has detected thorium, potassium, oxygen, magnesium, silicon, calcium, titanium, iron and *uranium* in the moon's crust since its launch in 2007.

take over transportation between earth space stations and the moon and planets.

The decided advantages of nuclear rockets in terms of speed and economy will eventually make them the preferred method of travel in the solar system.

A5 Mars Colonization

A5.1 Transportation to Mars

Although there are many ingenious proposals for travel to Mars and other planets the most economical, fastest and safe mode of transportation will be nuclear rockets after nuclear rocket technology has matured through use in space. There is no substitute for the practical experience, and the innovation that it stimulates, through the actual use of nuclear rockets. Examples of similar growth in the efficiency and sophistication of vehicles are the history of automobile and airplane development.

Nuclear rocket traffic to Mars would initially originate from moon orbit until rocket safety had reached a point where trips to Mars could originate from earth orbit.

Initially the Martian moons could be used as terminals for traffic from earth – with a shuttle service to the planet's surface.

A5.2 Changing the Atmosphere on Mars

The recent discovery of large amounts of water and frozen carbon dioxide on Mars gives hope to the possibility of transforming the Martian atmosphere into a thicker carbon dioxide atmosphere that eventually could evolve to have sufficient oxygen for human needs. There appear to be a number of feasible approaches. One approach might be to explode large hydrogen bombs to vaporize the Martian ice caps and create a thicker atmosphere with carbon dioxide and water vapor. Thinking on a 50 – 100 year time frame the effects of radioactivity would not be of importance. The possibility of a "nuclear winter" resulting from the explosion of the bombs could be reduced by exploding them below ground at a depth sufficient to minimize dust thrown into the atmosphere.

Another approach would be to divert icy comets or meteors using nuclear explosives to bombard Mars. Recently a reservoir of icy comets has been found circling Jupiter. In addition to adding water to the planet they would also serve to heat it.[67]

A5.3 Introduction of Plant life

After the generation of a relatively thick and warm atmosphere the introduction of (perhaps genetically modified) plants and eventually animals on a planetary scale would become feasible. The primarily carbon dioxide atmosphere could then be transformed to a breathable atmosphere containing oxygen so that humans could live in a manner similar to that of higher altitude regions of earth.

Then colonization could proceed with the development of business, industry and tourism. Humanity would then have another home world.

A6 Venus Colonization – A Long Term Project

A6.1 Proposal for a Tri-Planet Homeland for Humanity

The transformation of Venus into a home for mankind also may be possible over a period of thousands of years. In this case we must trap almost 99% of its (primarily) carbon dioxide atmosphere within the rock on the planet's surface creating an earth-like atmospheric pressure and reducing the greenhouse effect (perhaps through a nuclear winter or comet bombardment).

We must also remove sulfur from the atmosphere.[68] Next we must generate an atmosphere containing oxygen and surface water (possibly through bombardment with diverted icy comets as in the case of Mars). The remaining issues: the slow rotation of Venus and the lack

---

[67] Recently evidence has surfaced that Mars may have been bombarded by icy comets and meteors in the past that created massive water movements as well as heating the planet to earth-like temperatures for periods lasting up to hundreds of years. See the article: **Environmental Effects of Large Impacts on Mars** by Teresa L. Segura, Owen B. Toon, Anthony Colaprete, and Kevin Zahnle *Science* **298** (Dec. 6, 2002) pages 1977-1980.

[68] The early earth also had very little oxygen and enormous amounts of methane and sulfur in the atmosphere that was eventually trapped in rocks forming sulfites. See the articles: Farquhar, J., K.D. McKeegan and M.H. Thiemens, **Mass-independent sulfur of inclusions in diamond and sulfur recycling on early Earth** *Science* **298** (Dec. 20, 2002) pages 2369-2372; Habicht, K.S., and D.E. Canfield, **Callibration of Sulfate Levels in the Archean Ocean** *Science* **298** (Dec. 20, 2002) pages 2372-2374; Wiechert, U.H., **Earth's Early Atmosphere** *Science* **298** (Dec. 20, 2002) pages 2341-2342.

of a magnetic field create complications but they can also be overcome by an advanced society.

Some might feel that the investment of vast efforts over a period of perhaps thousands of years is not a reasonable proposal. However, the reward for such an effort would be another earth at a relatively close distance. Thus humanity would benefit from an enormous expansion of its living space. The accomplishment of this goal would also have the beneficial effect of vastly increasing the scope of human civilization.

A6.2 Interplanetary Civilization

One of the major points of the work of Arnold Toynbee, and other students of civilization, is that a civilization grows and matures through meeting great challenges. As the Egyptians conquered the Nile Valley to produce a great civilization, Mankind has the opportunity to conquer the nearby planets to achieve a new level of civilization.

The accomplishment of these projects will give humanity three earth-like homes. If large extraterrestrial human colonies are established then a human space civilization will develop. *The size of the challenge confronting this future civilization is such that a successful response will undoubtedly create a civilization that may be an order of magnitude above current human civilizations.* The mushroom ring of civilization will then have expanded to the planets with a clear view towards the stars.

The United States, the European nations, Russia, China and India are the only entities with sufficient resources to lead a major move into space. Rather than devote resources to war (and preparations for war) they should create a major presence in space.

Militarism will waste the resources of the world just as Trajan wasted the resources of Rome in the expansion of its frontiers. EuroRussoSinoIndianAmerican resources could be better devoted to another purpose: creating major space settlements—that would be to the world as the United States was to Great Britain—the offspring that helped to save the parent (in two world wars.)

A7 Solar System Colonization

Not content to colonize the inner planets it is reasonable to consider the development of settlements in the asteroid belt, the moons of the outer planets, and perhaps even as far as the Oort Belt.

Initially such settlements would be for scientific purposes. One could imagine, for example, an array of radio and optical telescopes ranging from one edge of the solar system to the other. If the components of the array were properly synchronized as we do in telescope arrays on earth at present we would have a giant eye on the universe that would help us detect cosmic phenomena, and particularly, detect solar systems around other stars with earthlike planets. This telescope array would then help determine the target stars to visit as we begin to explore the stars.

Eventually colonies in the outer solar system might be developed for mining, industry and other purposes. The low intensity of sunlight in those regions would require artificial light for health and other reasons.

REFERENCES

Blaha, S., 2004, *Quantum Big Bang Cosmology: Complex Space-time General Relativity, Quantum Coordinates, Dodecahedral Universe, Inflation, and New Spin 0, ½, 1 & 2 Tachyons & Imagyons* (Pingree-Hill Publishing, Auburn, NH, 2004).

_____, 2006, *A Unified Quantitative Theory Of Civilizations and Societies: 9600 BC - 2100 AD* (Pingree-Hill Publishing, Auburn, NH, 2006)

_____, 2007a, *Physics Beyond the Light Barrier: The Source of Parity Violation, Tachyons, and A Derivation of Standard Model Features* (Pingree-Hill Publishing, Auburn, NH, 2007).

_____, 2007b, *The Origin of the Standard Model: The Genesis of Four Quark and Lepton Species, Parity Violation, the ElectroWeak Sector, Color SU(3), Three Visible Generations of Fermions, and One Generation of Dark Matter with Dark Energy* (Pingree-Hill Publishing, Auburn, NH, 2007).

_____, 2008, *A Complete Derivation of the Form of the Standard Model With a New Method to Generate Particle Masses SECOND EDITION* (Pingree-Hill Publishing, Auburn, NH, 2008)

_____, 2009a, *Bright Stars, Bright Universe* (Pingree-Hill Publishing, Auburn, NH, 2009)

_____, 2009b, *To Far Stars and Galaxies: Second Edition of Bright Stars, Bright Universe* (Pingree-Hill Publishing, Auburn, NH, 2009).

_____, 2010a, *The Standard Model's Form Derived from Operator Logic, Superluminal Transformations and GL(16)* (Pingree-Hill Publishing, Auburn, NH, 2010).

_____, 2010b, *SuperCivilizations: Civilizations as Superorganisms* (McMann-Fisher Publishing, Auburn, NH, 2010).

Freeman, Marsha, *Krafft Ehricke's Extraterrestrial Imperative* (Apogee Books, 2009).

Huang, K., 1992, *Quarks, Leptons & Gauge Fields Second Edition* (World Scientific, River Edge, NJ, 1992).

Huang, K., 1998, *Quantum Field Theory* (John Wiley, New York, 1998).

Lee, S. Y., 2004, *Accelerator Physics Second Edition* (World Scientific Publishing co, New Jersey, 2004).

Mallove, E. F. and Matloff, G. L., (1989) *The Starflight Handbook* ((John Wiley, New York, 1989).

Schmidt, S. and Zubrin, R. (eds.), 1996, *Islands in the Sky* (John Wiley, New York, 1996).

Weinberg, S., 1972, *Gravitation and Cosmology* (Wiley, New York, 1972).

Zubrin, R., 2000, *Entering Space* (Penguin Putnam, New York, 2000).

About the Author

Stephen Blaha is an internationally known physicist with extensive interests in Science, the Arts, and Technology. He received his Ph.D. in Theoretical Physics from Rockefeller University (NY). He has written a highly regarded book on physics, consciousness and philosophy – *Cosmos and Consciousness*, a book on Science and Religion entitled *The Reluctant Prophets*, a book applying physics concepts to the history of civilizations, and books on Java and C++ programming. He developed a mathematical theory of civilizations that is described in *The Life Cycle of Civilizations*. Recently he completed a major new study of Cosmology: *Quantum Big Bang Cosmology: Complex Space-time General Relativity, Quantum Coordinates, Dodecahedral Universe, Inflation, and New Spin 0, ½, 1 & 2 Tachyons & Imagyons*. He has served on the faculties of several major universities. He was an Associate of the Harvard Physics Faculty for twenty years (1983-2003). He was also a Member of the Technical Staff at Bell Laboratories, a member of management at the Boston Globe Newspaper, a Director at Wang Laboratories, and President of Blaha Software Inc and Janus Associates Inc. (NH).

Among other achievements he was a co-discoverer of the "r potential" for heavy quark binding developing the first (and still the only demonstrable) non-abelian gauge theory with an "r" potential; first suggested the existence of topological structures in superfluid He-3; first proposed Yang-Mills theories would appear in condensed matter phenomena with non-scalar order parameters; first developed a grammar-based formalism for quantum computers and applied it to elementary particle theories; first developed a new form of quantum field theory without divergences (thus solving a major 60 year old problem that enabled a unified theory of the Standard Model and Quantum Gravity without divergences to be developed); first developed a formulation of complex General Relativity based on analytic continuation from real space-time; first developed a generalized non-homogeneous Robertson-Walker metric that enabled a quantum theory of the Big Bang to be developed without singularities at $t = 0$; first generalized Cauchy's theorem and Gauss' theorem to complex curved multi-dimensional spaces; first developed a physically acceptable theory of faster-than-light particles – tachyons – of any spin; first showed a universe with three complex spatial dimensions has an icosahedral symmetry; first developed the form of the composition of extrema in the Calculus of Variations; first quantitatively suggested that inflationary periods in the history of the universe were not needed; first proved Gödel's Theorem implies Nature must be quantum, first derived the form of the Standard Model, first showed how to resolve logical paradoxes including Gödel's Undecidability Theorem by developing Operator Logic and Quantum operator Logic, first developed a quantitative harmonic oscillator-like model of the life cycle, and interactions, of civilizations, and first developed an axiomatic derivation of the forms of The Standard Model with WIMPs from geometry – space-time properties.

Blaha was also a pioneer in the development of UNIX for financial and scientific applications, in financial modelling software, in database benchmarking, in networking (1982), in the development of Desktop Publishing (1980's), and in the development a hybrid shell programming technique (1982) that was a precursor to the PERL programming language. He received Honorable Mention in the Gravity Research Foundation Essay Competition in 1978, and was nominated for three "Awards for Technical Excellence" in 1987 by PC Magazine for PC software products that he designed and developed.